6th Agria Conference on Innovative Vehicle Technologies and Automation Solutions (InnoVeTAS 2022)

IOP Conference Series: Materials Science and Engineering
Volume 1237

Eger, Hungary
13 May 2022

Editors:

Laszlo Pokoradi
Laszlo Soltesz

ISBN: 978-1-7138-5438-8
ISSN: 1757-8981

Printed from e-media with permission by:

Curran Associates, Inc.
57 Morehouse Lane
Red Hook, NY 12571

Some format issues inherent in the e-media version may also appear in this print version.

Institute of Physics
Dirac House, Temple Back
Bristol BS1 6BE UK

Phone: 44 1 17 929 7481
Fax: 44 1 17 920 0979

techtracking@iop.org

Additional copies of this publication are available from:

Curran Associates, Inc.
57 Morehouse Lane
Red Hook, NY 12571 USA
Phone: 845-758-0400
Fax: 845-758-2633
Email: curran@proceedings.com
Web: www.proceedings.com

TABLE OF CONTENTS

The Sixth Agria Conference on Innovative Vehicle Technologies and Automation Solutions, InnoVeTAS2022

László Soltész[1]*, László Pokorádi[2], István Gödri[3]

[1] Emerson Automation FCP Ltd, H- 3300 Eger, Hungary
[2] Óbuda University, Institute of Mechatronics and Vehicle Engineering, Népszinház u. 8, H-1081 Budapest, Hungary
[3] Emerson Automation FCP Ltd, H- 3300 Eger, Hungary

*Email: laszlo.soltesz@emerson.com

Even though the last few years were filled with all kinds of challenges for companies and universities; we are happy to host once again, the Sixth Agria Conference on Innovative Vehicle Technologies and Automation Solutions, also known as InnoVeTAS2022. This conference series was started as a subset event of the International Pneumobile competition, in order to make it more attractive to students and professors from a wide range of universities in Europe. Today, the InnoVeTAS conference has become an independent scientific forum, which is traditionally held in the middle of May. The International Pneumobile conference date has been moved to September.

This being the sixth conference in the series; our event has become one of the major technical and scientific talks in Heves County! Additionally, we are now visible on the venue of Hungarian national conferences! In cooperation with IOP Publishing, this year's conference proceedings will also be the second time, that it will have been listed in the Scopus index, which is useful for high level researchers.

This conference's main theme is: To "Sense the future" - to focus on the future. During the last decades, we have seen a dramatically changed world regarding customer needs, needs for new products, and the needs for speed of new products to market. All in all, the world is speeding up. People are living their life with greater expectations. These changes require a speedy approach to the whole product life cycle, from market research, through production and delivery, via product development. In parallel to these sophisticated requirements, C-suite business executives continuously focus on increasing efficiency and profit margin. Because of these requirements, along with the needs for the reduction of stock levels; business customers, – not only in automotive areas, – started to require smaller order quantities. Due to its high-speed production and delivery capabilities, Emerson can provide the necessary components in an almost "just in time" fashion to industrial customers, according to their demand on new investment, repair activities or planned maintenance.

Business players were thinking that they were agile enough to manage the high speed of the market in an nimble way, but the global COVID-19 pandemic generated additional factors into the world of business. The world was immediately faced with special circumstances. The first challenge was, how to protect employees, increase the availability of safety products, and expand the infrastructure for home office, etc. In the next step, supply chain topics caused further issues, which had to be solved. Clearly this supply chain topic has only been partially addressed. Unfortunately, this year the war in Ukraine has made this situation even more problematic. Supply chain issues are expected to haunt the global economy for the next few years, perhaps even decades.

Nowadays, the world is trying to get back to a "normal" life, which has become really different since 2019. For example, these days, home office and hybrid work have become the norm at market leading companies; there is no questioning it anymore, employees expect it. Universities have also changed their undertakings on how to organize education related activities, exams, and project work. Emerson, with its high level of support for STEM endeavors while providing various scholarship programs for many different students, clearly shows the importance of investment into future! The sixth Agria Conference on Innovative Vehicle Technologies and Automation Solutions, InnoVeTAS2022 has been organized in hybrid way. With no more restrictions to live gatherings, hopefully we will have had a strong, live, face to face, presence by professors, researchers, students, and interested people. Personal discussions and meetings are, just as an important part of the event, as professional presentations, and the publication of papers.

We would like to thank all of the authors for contributing research papers. Without these efforts by the experts, there would have been no conference at all. Additionally, we would like to thank all of the people who have been supportive of the conference, have helped, and have encouraged us in all activities related to this symposium.

Organizing Committee

- László Soltész (chair)
- Zsófia Kádár
- Mihály Varga
- Sándor Kiss
- Judit Fehér
- Szilárd Nagy

Scientific Committee

- László Pokorádi (founder chair)
- L. Ambroziak, Bialystok University of Technology
- L. Berényi, University of Miskolc
- J. Bihari, University of Miskolc
- Zs. Farkas, Budapest University of Technology and Economics
- L. Fechete, Technical University of Cluj-Napoca
- W. Fiebig, Wroclaw University of Science and Technology
- D. Fodor, University of Pannonia
- Z. Forgó, Sapientia Hungarian University of Transylvania
- L. Földi, Szent István Egyetem
- R. Jánoš, Technical University of Košice
- Gy. Juhász, University of Debrecen
- G. Kátai-Urbán, John von Neumann University
- M. Madissoo, Estonian University of Life Sciences
- P. Mazeika, Klaipeda University
- P. Megyeri, University of Pécs

- Gh. Mihai, University Politehnica of Bucharest
- L. Osiecki, Gdańsk University of Technology
- V. Pirs, Latvian University of Agriculture
- V. Priede, Riga Technical University
- K. Psiuk, Silesian University of Technology
- M. Ranuša, Brno University of Technology
- M. Simon, Universitatea Petru Maior
- T. Szakács, Óbuda University
- L. Soltész, Emerson Automation FCP Kft.
- T. Szabó Budapest University of Technology and Economics
- T.I. Tóth University of Szeged
- E. Tóth-Laufer, Óbuda University
- T. Mitran, University of Oradea
- I. Péter-Szabó, University of Szeged, Faculty of Engineering

Editorial Committee

- László Pokorádi
- László Soltész

Invited Speakers

- István Gödri, Site and Operations Director, Emerson Automation FCP Ltd, Hungary
- Jerome Thomas, Vice President FLMC Operations EMEA

Organizing Institution

- Óbuda University, Institute of Mechatronics and Vehicle Engineering

Sponsor

- Emerson Automation FCP Ltd.

IOP Conf. Series: Materials Science and Engineering 1237 (2022) 011002 doi:10.1088/1757-899X/1237/1/011002

Peer review declaration for InnoVeTAS2022

All conference organisers/editors are required to declare details about their peer review. Therefore, please provide the following information:

- **Type of peer review:** *Single-blind*
- **Conference submission management system:** *none*
- **Number of submissions received:** *25*
- **Number of submissions sent for review:** *21*
- **Number of submissions accepted:** *20*
- **Acceptance Rate (Number of Submissions Accepted / Number of Submissions Received X 100):** *80*
- **Average number of reviews per paper:** *2.0*
- **Total number of reviewers involved:** *31*
- **Any additional info on review process:** *none*
- **Contact person for queries (please include: name, affiliation, institutional email address)**
  László Pokorádi; Óbuda University; pokoradi.laszlo@bgk.uni-obuda.hu;
  László Soltész; Emerson Automation FCP Ltd..; laszlo.soltesz@emerson.com

Please submit this form along with the rest of your files on the submission date written in your publishing agreement.

The information you provide will be published as part of your proceedings.

Optimization of compressed air consumption in a pneumatic vehicle

P Marko and W Fiebig

Wroclaw University of Science and Technology, Faculty of Mechanical Engineering, Wybrzeże Wyspiańskiego 27, Wrocław, Poland

Email: piotr.marko@pwr.edu.pl

Abstract. The aim of the research described in the article was to optimize compressed air consumption in a pneumatic driven vehicle to achieve the maximum traveled distance and maximum acceleration over a short distance. The Pneumatic vehicle was equipped with a 10-liter tank. The vehicle drive system consists of an actuator, a reciprocating movement which is converted into rotary motion and transmitted to the road wheels. The article analyses the impact on the range and maximum speed of a pneumatic vehicle based on parameters such as gear ratio, the amount of air supplied to the actuator, opening the drain valves before the piston reaches dead center, setting on the reduction valve, size of the buffer tank, and gear shift time. After the initial analytical estimation of the actuator operating parameters, a numerical analysis of the vehicle operation of the pneumatic and drive systems of the vehicle was performed in the Simcenter Amesim program. In this program, a pneumatic system model was created together with the vehicle drive system. Finally, conclusions from the simulations are presented, showing how the pneumatic system should be built and how it should be powered to obtain the maximum range and the greatest acceleration of the vehicle.

1. Introduction

Nowadays, alternative ways of powering vehicles are being searched for. One possibility is the pneumatic drive, which was used in cars and trams from the end of the nineteenth century to the beginning of the twentieth century. This type of vehicle propulsion was replaced by internal combustion engines. Since the 1970s, thanks to various engineers from all over the world, this drive has returned to favor, and by that time many compressed air vehicles had already been created [1].

The main problem in designing a compressed air vehicle is the low energy content of compressed air, which is 0.5 MJ / kg and 0.2 MJ / l [2]. For comparison, the energy density of gasoline is 46.4 MJ / kg and 34.2 MJ / l, LPG 46.4 MJ / kg and 26 MJ / kg, diesel oil 45.6 MJ / kg and 38.6 MJ / l [2, 3]. To compete with other types of drive, the key aspect of compressed air vehicles is to achieve the highest possible efficiency to maximize the use of the already small amount of energy in compressed air. The efficiency of pneumatic systems is low if they do not use any energy recovery methods or optimize compressed air consumption. The efficiency of pneumatic tools is in the range of 10-15% [4].

Yusop [5] dealt with optimizing compressed air consumption. The author gives some advice on how to reduce compressed air consumption: the piston should reach the end of the cylinder; inlet valves should be mounted on cylinders with as short pipes as possible and use short-stroke cylinders with appropriately selected mechanical adjustment instead of long-stroke cylinders.

One way to recover compressed gas consumption is to use a flexible tank that can be charged with air flowing from the actuator. This type of energy recuperation has been used in ankle and foot orthosis [6]. The authors used 25% less gas in their system than in the system without recuperation.

This type of recovery was also investigated by Cummins [7] et al. Scientists initially expanded the air in a small-bore pneumatic cylinder, then the air was injected into a flexible reservoir to finally be released into a large-bore pneumatic cylinder. Researchers boast of increasing air consumption efficiency by up to 72%.

Another possibility to increase the efficiency of compressed air consumption is to use a bypass valve. Yang [8] et al. compared the compressed air consumption of an actuator powered by a 5/3 valve with the compressed air consumption of a cylinder; when the actuator chambers were additionally connected by a 3/2 valve, which allowed air to flow from the chamber with a piston rod to the chamber without a piston rod. They concluded that the use of a bypass valve allows compressed air energy savings of 12% to 28% compared to a system without a bypass valve.

Szakacs [9] was involved in the construction of a block model of the pneumatic system of a pneumatic vehicle in the MATLAB program. The author assigned each block to physical equations that describe the behavior of the gas in particular parts of the system. After the simulation, it was possible to observe the course of the system parameters, which allows testing the consumption of compressed air at various settings.

2. Research description

The study investigated various ways of powering the actuator to use the compressed air stored in the tank more effectively. The reciprocating movement of the actuator was converted into the movement of the pneumatic vehicle.

The aim of the work was to determine the input parameters of the pneumatic system, which:

- ensure the maximum possible range of the vehicle at an average speed of at least 15 km / h and,
- ensure the maximum speed of covering a distance of 250 m.

A diagram of the vehicle's pneumatic system is shown in Figure 1.

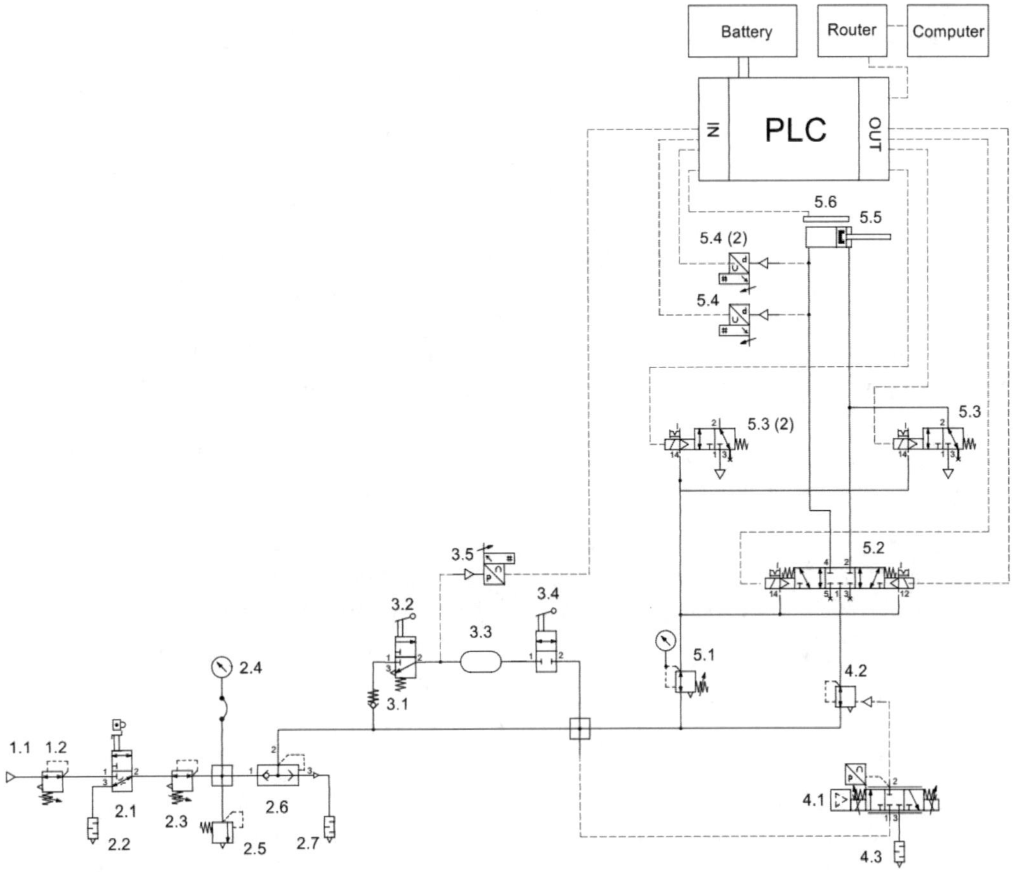

Figure 1. Diagram of the pneumatic system.

The study investigated the following ideas for the maximum use of energy stored in compressed air:

1. Partial power supply to the actuator. The goal is to make the actuator move through the expanding gas and not through the air flow.
2. Various values of gear ratios; the values of gear ratios directly depend on the values of the actuator resistances.
3. Minimizing pressure reduction on the reduction valve (1.2) in the pressure reservoir (1.1) – various values of the pressure supplying the actuator were tested.
4. Using buffer tanks (3.3) of different sizes – the tank should be large enough to be able to supply the actuator with air dynamically and collect heat from the environment to heat the air (the temperature of air has decreased after the reduction), but at the same time small enough so that the air does not expand unnecessarily in it.

The following ideas for maximizing vehicle speed were investigated in the study:

1. Deflating air before reaching dead center and simultaneously energizing the opposite chamber, this is to accelerate the piston travel through dead center.
2. Partial power supply of the actuator; this is aimed at a more efficient use of compressed air from the buffer tank (the air supply from the pressure tank is strongly limited by the reducer).

3. Using buffer tanks of different sizes, this is to store a large amount of compressed air downstream of the regulator, through which there is a small flow.
4. Gear shift time - This is to obtain a high torque output from the variator at the beginning of travel in order to overcome inertial resistance. As the vehicle speed increases, the variator will reduce the torque and increase the rotational speed on the transmission output shaft.

The designed circuit works as follows. Air from a 10-litre compressed air container (1.1) with a working pressure of 200 bar is reduced at the pressure reducing valve (1.2) to pressures ranging from 3 bar to 10 bar. Then the air flows through the safety stop circuit (2.1, 2.2, 2.3, 2.4, 2.5, 2.6, 2.7). Later, the air flows through the buffer tank (3.3) and other valves that enable its connection (3.1, 3.2, 3.4). The pressure regulator (4.2) then supplies air to the solenoid valve (5.2). Pressure regulator 4.2 is controlled by a pressure regulator (4.1), which is electrically controlled in proportion to the displacement of the accelerator pedal.

The solenoid valve (5.2) directs air to the right or left actuator chamber (5.5). When the piston moves to the right, air from the right chamber is released by valve (5.3), and in the event of left-hand movement of the actuator, air from the left chamber is discharged through the valve (5.3 (2)). The PLC controls the opening and closing of the valves. It operates based on information about the position of the piston from the position sensor (5.6). The pneumatic actuator drives the drive train, which transfers power to the wheels.

The controller is powered by two 12 V batteries connected in series. Data from pressure sensors (3.5, 5.4, 5.4 (2)) and from the piston position sensor (5.6) are transmitted through the controller and router to the computer, so that the pressure waveforms and the piston position can be analyzed for the correct efficiency of system operation.

In the Simcenter Amesim program, a block model was built containing all pneumatic and mechanical elements of the system, as well as control functions to be performed by a PLC controller.

The block model representing the pneumatic system is shown in Figure 2. Each block has boundary values such as pressure, flow rate coefficient, orifice area, etc.

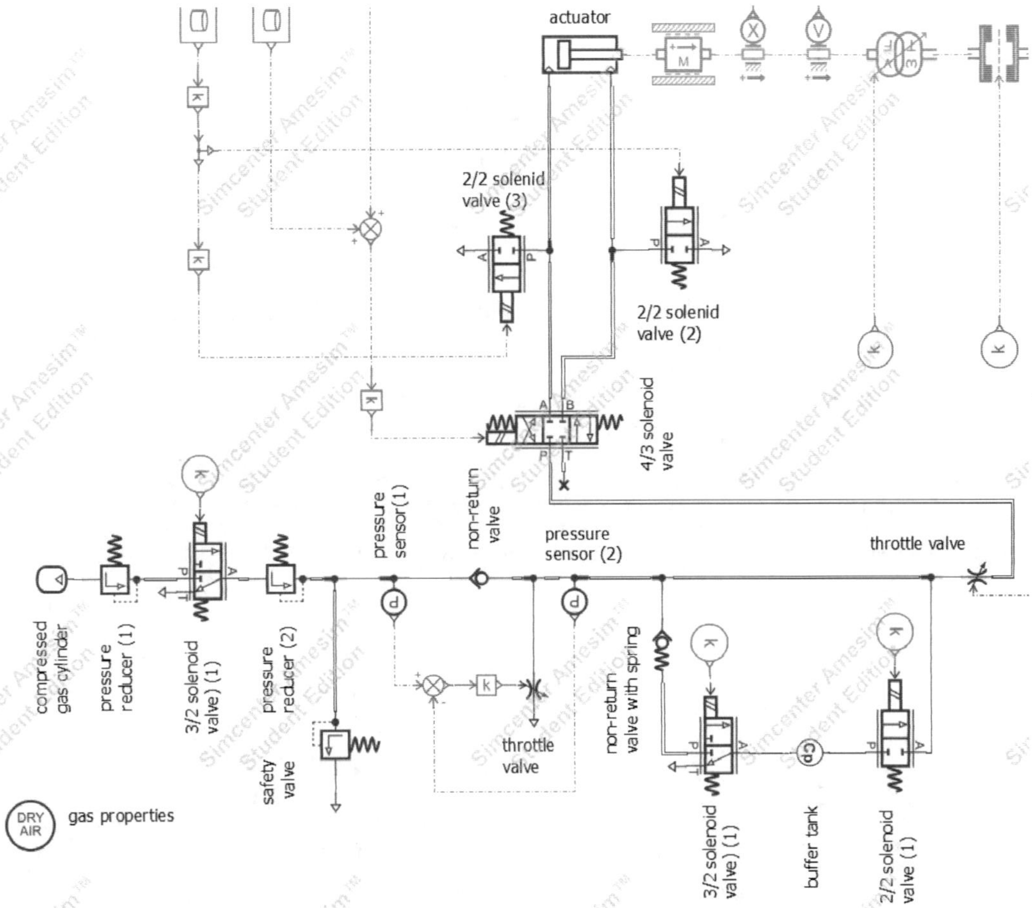

Figure 2. Pneumatic block diagram (Simcenter Amesim).

3. Measurement results

3.1. Economic Driving

Measurements were carried out for five variants of the size of the supplied chamber: 10, 20, 40, 60 and 100%. For example, the variant of 10% power supply to the chamber consisted of feeding the chamber while the piston was 5 cm from its dead center (the stroke of the actuator is 50 cm). After exceeding 5 cm, the air supply was cut off. The further movement of the piston occurred due to the expansion of the gas.

In each variant, the vehicle parameters were tested for four pressure values (4, 6, 8, 10 bar) and four gear ratio values in the variator in the drive system (0.5, 1, 1.5, 1.9).

The diagrams show the range of the vehicle as a function of the setting of the reducing valve for different values of the gear ratio. There is a different graph for each air dose.

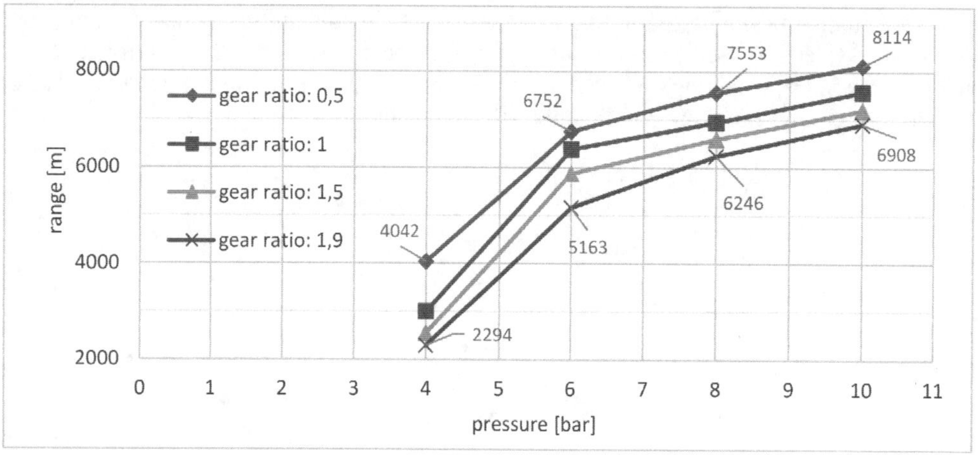

Figure 3. Range-pressure graph (the amount of air dose is 10%).

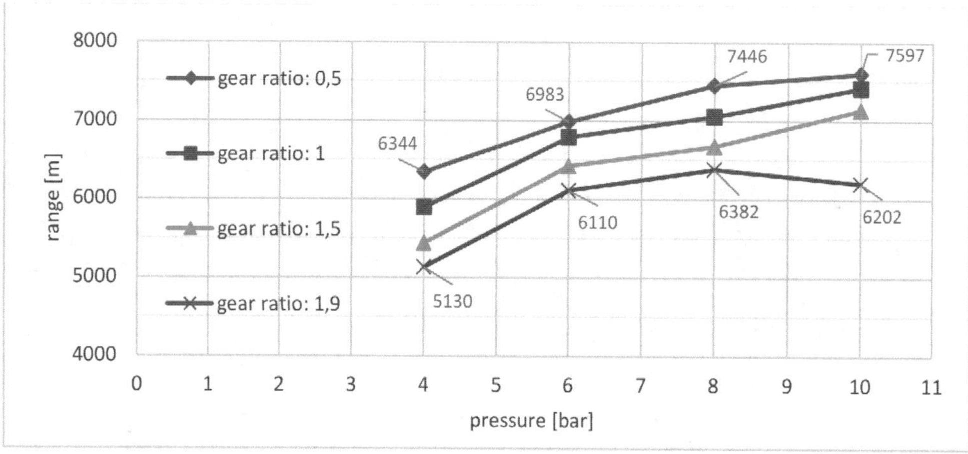

Figure 4. Range-pressure graph (the amount of air dose is 20%).

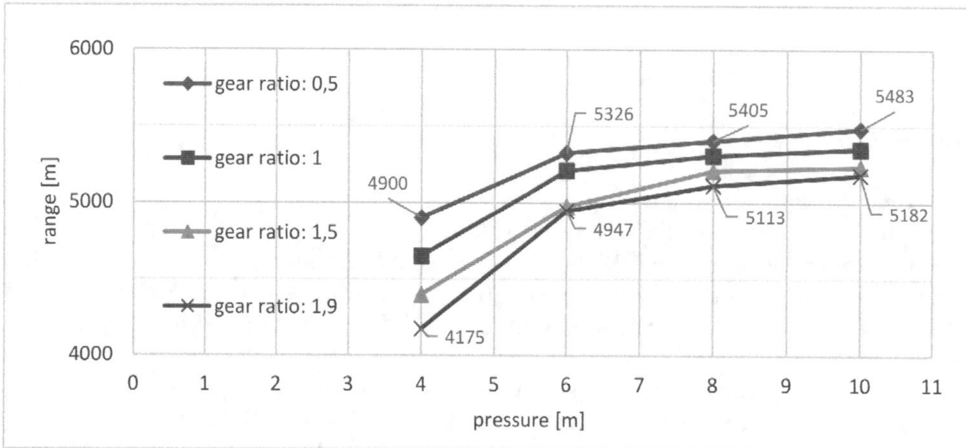

Figure 5. Range-pressure graph (the amount of air dose is 60%).

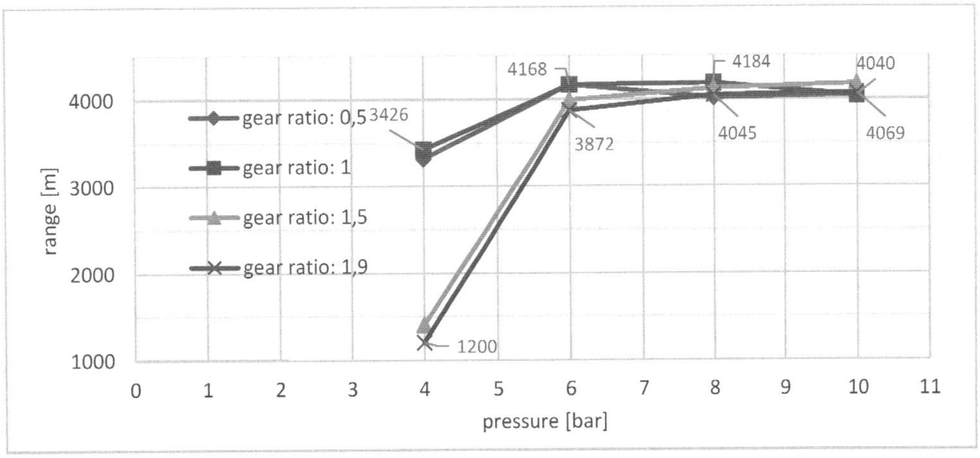

Figure 6. Range-pressure graph (the amount of air dose is 100%).

In most cases, the vehicle traveled further, with a lower dose of air supplied to the actuator; the higher the pressure setting was on the reduction valve, the lower was the gear ratio. It was noticed that the vehicle had the greatest range (8114 m) for the air dose of 10%, the setting of 10 bar on the reduction valve and for the smallest gear ratio (0,5 – maximum multiplication of gear rotational speed, maximum load on the piston rod). For this air supply variant, figure 7 shows a diagram showing the position of the piston and the pressure in the cylinder chambers as a function of time. In the chart, we can notice the delivery of a dose of compressed air to the chambers of the actuator and the next expansion which causes the piston to move.

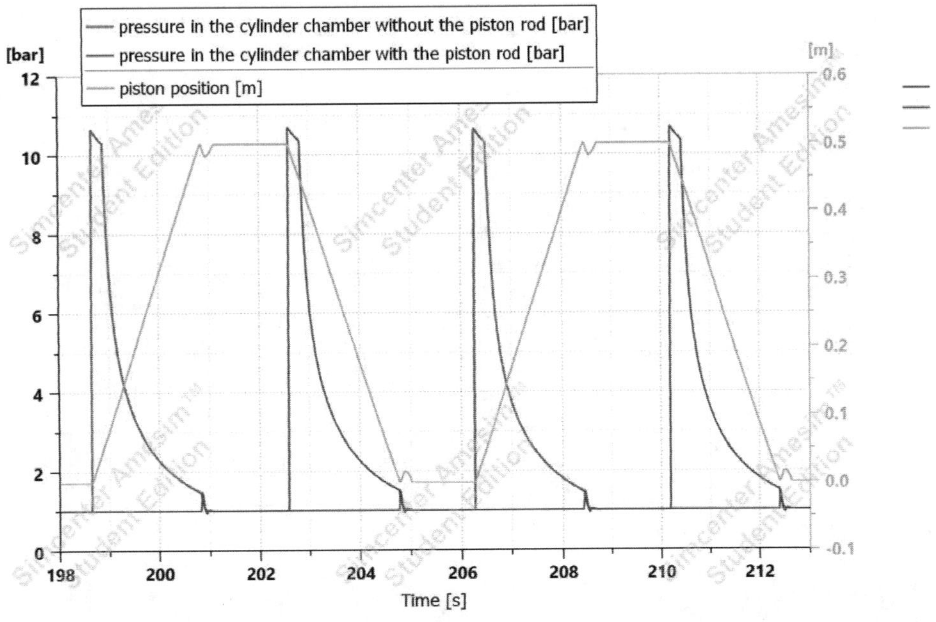

Figure 7. Diagram of the displacement of the piston and the pressure in the cylinder chambers as a function of time for the amount of air dose equal to 10%.

For these parameters of the system, further studies were carried out to select the optimal size of a buffer tank. Vehicle range tests were carried out for 5 buffer tank sizes (0 l, 1 l, 5 l, 14 l, 24 l).

The test results are shown in diagram 8.

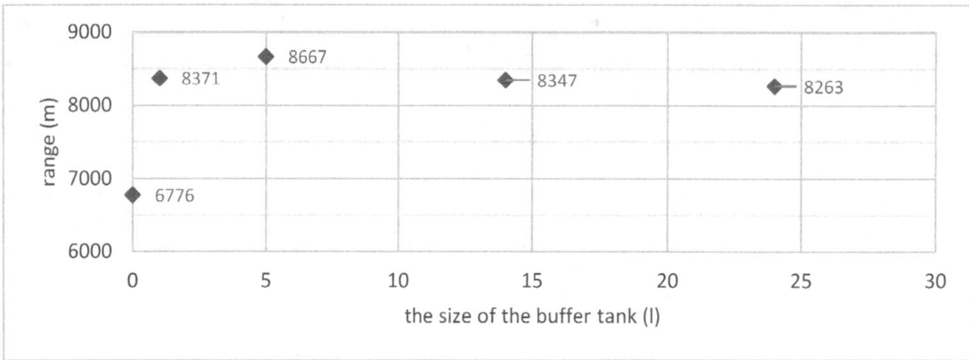

Figure 8. Influence of the size of the buffer tank on the range of the vehicle.

The diagram shows the range of the vehicle depending on the size of the buffer tank. The vehicle has traveled the longest distance for a 5-l tank (8667 m). The range was shorter with the use of tanks with a capacity of 1 l (8371 m), 14 l (8347 l), and 24 l (8263 m). Complete abandonment of the tank significantly reduces the range of the vehicle.

The size of the buffer tank has little effect on the efficiency of air consumption. However, it cannot be completely dispensed with, because then it is not possible to supply the actuator chamber with a dose of air at a pressure of 10 bar, even with an air dose of 10%. Consequently, the cylinder chambers are supplied with a lower pressure.

Figure 9 shows the maximum range of the vehicle obtained during the simulation and the pressure drop in the pressure tank (1.1 in Figure 1) as a function of time for the most optimal system parameters for economic driving.

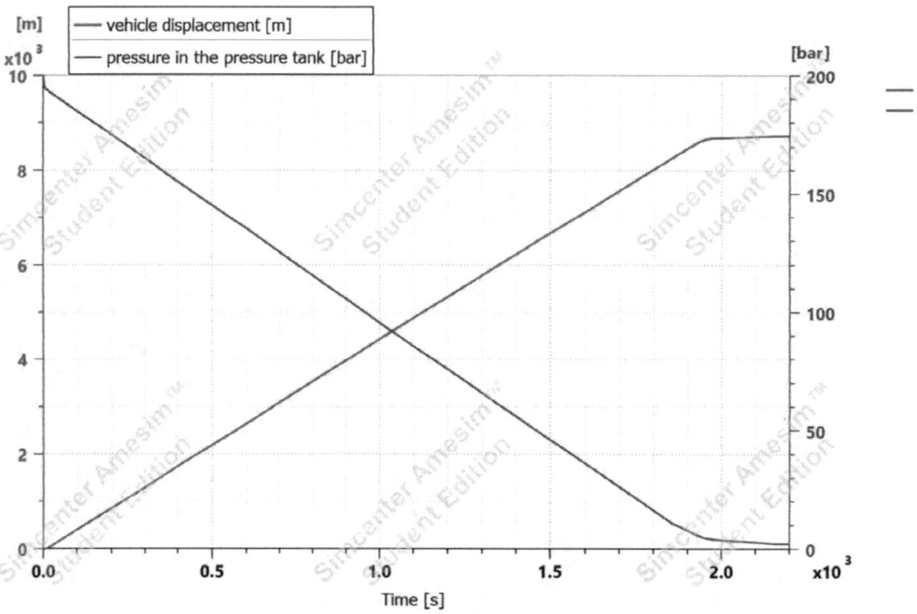

Figure 9. Graph of the greatest range and the pressure drop in the pressure tank as a function of time.

3.2. *Driving fast*

To maximize the vehicle speed and minimize the driving time of 250 meters, the following parameters were tested:

- the size of the air supply to the actuator chamber [%],
- early opening of the intake and drain valves before reaching the dead position [cm],
- gear shift time (from the variator gear ratio of 1.9 to the gear ratio of 0.5) [s],
- the size of the buffer tank [l].

The results in the form of graphs are shown in Figure 10.

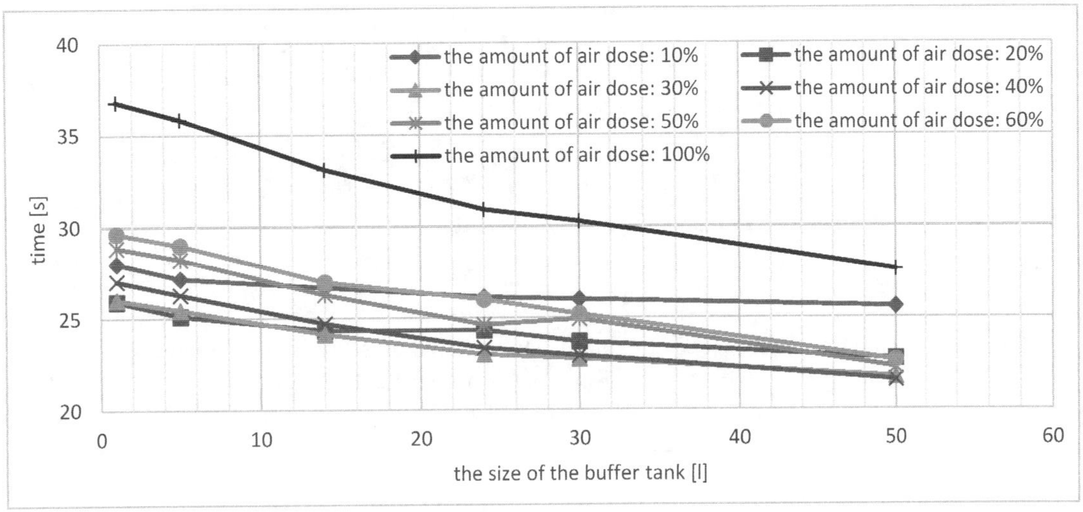

Figure 10. Graph of the travel time of 250 m as a function of the size of the buffer tank for various amounts of air dose.

The travel time of 250 m is shorter, the larger the buffer tank. The travel time of the 250 m section decreases with decreasing the value of the air dose to 40% and then increases.

The pressure diagrams (figures 11 and 12) are shown to explain the different 250 m travel times for different air doses. They show the pressure in the cylinder chamber with the piston rod, the pressure in the cylinder chamber without the piston rod, and the pressure in the buffer tank as a function of time for two extreme values of air doses: 10% and 100%.

The graph in Figure 11 shows the pressure in the buffer tank and the pressure in the actuator chambers as a function of time for the air dose of 10%. After driving a distance of 250 m, the pressure in the buffer tank is approximately 6.5 bar (absolute pressure).

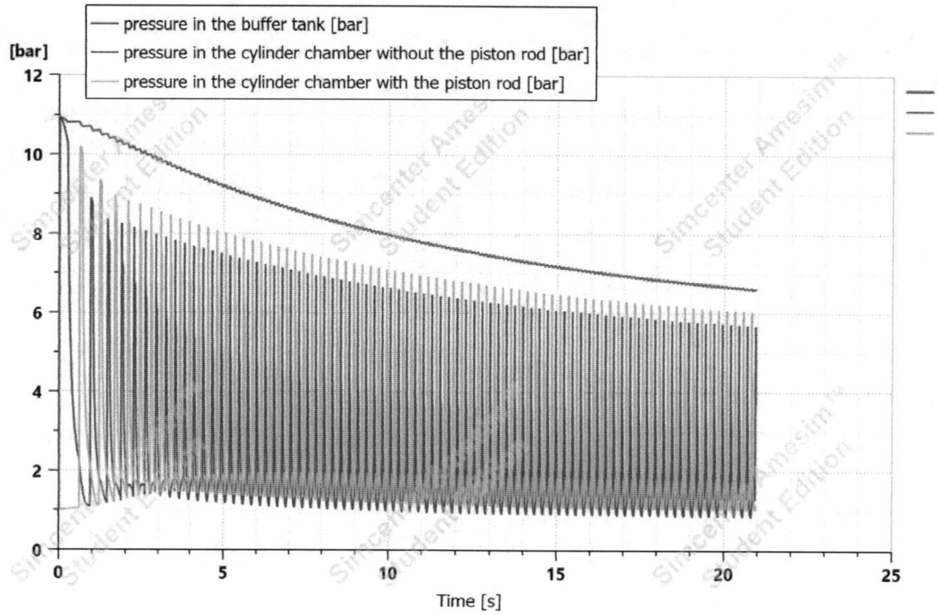

Figure 11. Graph of the pressure in the buffer tank and the pressure in the actuator chambers as a function of time for the air dose of 10%.

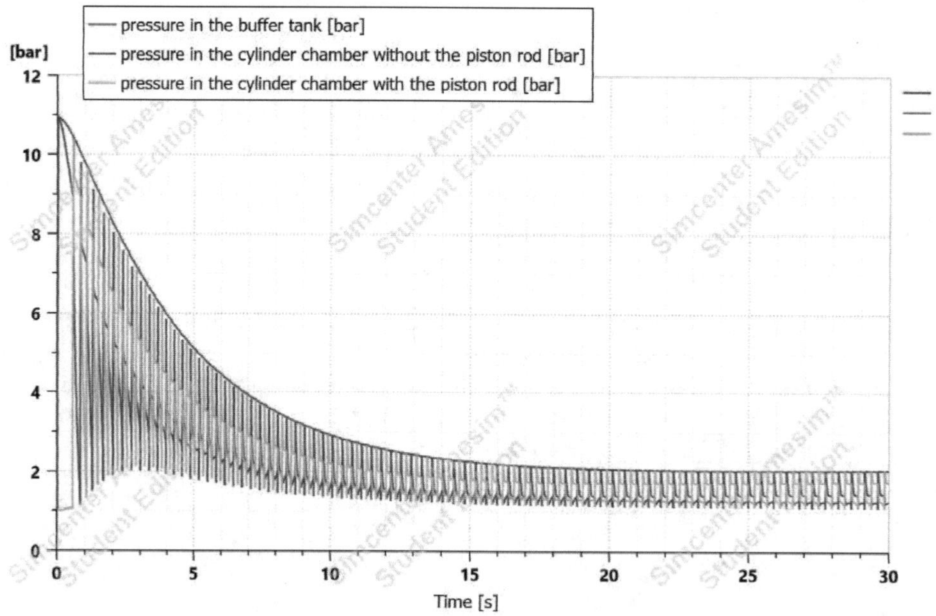

Figure 12. Graph of the pressure in the buffer tank and the pressure in the cylinder chambers as a function of time for the air dose of 100%.

The diagram in Figure 12 shows the pressure in the buffer tank and the pressure in the actuator chambers as a function of time for the air dose of 100%. After driving a distance of 250 m, the pressure in the buffer tank is approximately 2 bar (absolute pressure). This graph shows a more dynamic pressure drop in the buffer tank compared to the variant of feeding the actuator with a 10% air dose. The pressure drop in the buffer tank results from the greater demand of the actuator for air than the flow through the pressure reducer (block 1.2 in Figure 1).

For the variant for which the vehicle traveled 250 meters in the shortest time, further optimization simulations were carried out. The effects of early valve opening and shifting timing on vehicle travel of 250 m were measured. The simulation results are shown in the diagram in Figure 13.

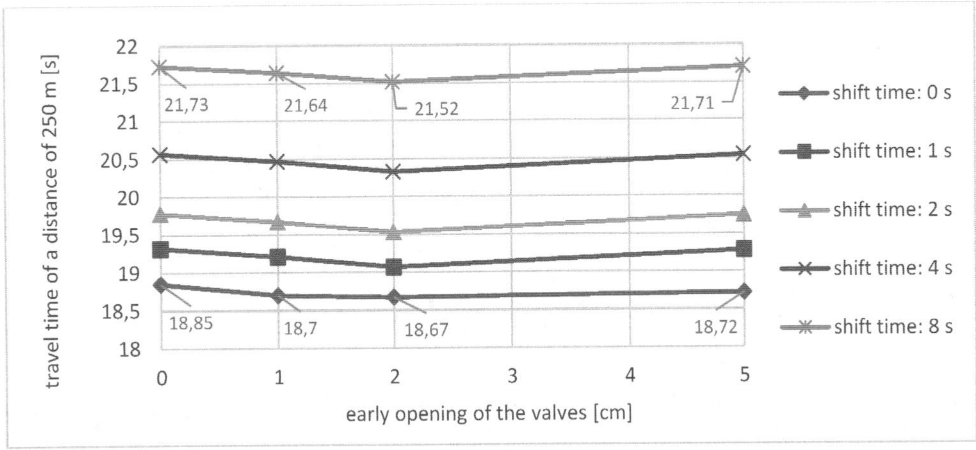

Figure 13. Graph of the 250 m travel time as a function of earlier valve opening for different times gear overrides.

The graph shows that the shorter the shift time was, the shorter was the travel time of 250 meters. This parameter is very important – between the travel time of 250 m for the gear shift time of 0 s, and the gear travel time of 8 s is a difference of more than 2.5 s. From this analysis it follows that the variator used in the vehicle is not needed. The actuator can generate a high torque on the drive wheel from the start of the vehicle motion.

The travel time of 250 m decreases with the increase of the earlier opening of the valves to the value of 2 cm, and after exceeding this value it increases.

Figure 14 shows the velocity and the greatest range of the vehicle as a function of time for the most optimal parameters of the pneumatic system for driving fast.

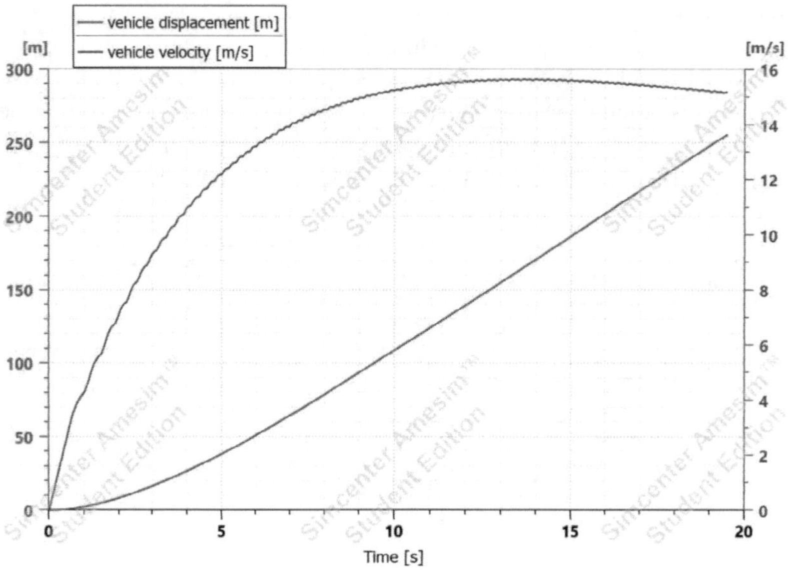

Figure 14. Graph of the velocity and the greatest range of the vehicle as a function of time.

4. Summary

Thanks to the research, it was possible to indicate in which areas to look for optimal driving parameters (actuator operation). There are the following measures to minimize air consumption:

- Reduce the pressure drop on the reduction valve as much as possible - supply the actuator with doses of air at a pressure of 10 bar, which is the maximum pressure at which the pneumatic elements used in the system can operate,
- Adjust the energy of the air dose delivered to the actuator to the work so that the actuator will have to perform. This can be done by:
    - supplying brief air to the actuator chamber and then cutting off the air supply so that the piston moves as a result of air expansion,
    - appropriate selection of the gear ratio in the drive transmission system in such a way that the resistance on the piston rod is properly matched to the force generated by the expanding air,
- Select the pressure and amount of the air dose as well as the transmission ratio in the drive system in such a way that the piston is always able to move the entire stroke without stopping between dead centers,
- Select a buffer tank with a minimum capacity that will allow for dynamic supply of the actuator chamber with high pressure air.

There are the following measures to maximize vehicle speed:

- Use a buffer tank with a maximum capacity (during the dynamic supply of the actuator, the demand for air by the actuator is greater than it can flow through the reducing valve),
- Adjust the amount of air delivered to the actuator. It must be selected in such a way as to optimally use the air from the buffer tank. With a too high dose of air supplied to the actuator from the buffer tank, the air will be used very quickly and the air from the cylinder will not be able to flow through the reducing valve rapidly enough. This will result in feeding the actuator with low-pressure air doses. If the dose of air is too low, the air from the buffer tank will not be fully used and after driving a distance of 250 meters, the tank will still contain high-pressure air,

- Start deflating the air expansion chamber before the piston reaches its dead center and simultaneously start supplying the opposite actuator chamber. This allows the piston to reverse the direction quickly when it is in the dead center,
- Reduce the pressure drop on the reduction valve as much as possible – supply the actuator with doses of air at a pressure of 10 bar, which is the maximum pressure at which the pneumatic elements used in the system can operate,
- When the cylinder is used to carry a heavy load from the very beginning of its operation, the cylinder can generate a large force on the piston rod right away. The use of a variator to maximize torque has proved redundant in this case.

5. Conclusions

The article proposes the possibility of powering the actuator in such a way that the pneumatic vehicle has the greatest range and covers a distance of 250 meters in the shortest time. In both cases, the most important parameter turned out to be the amount of air delivered to the actuator.

In the case of economic driving, an equally important parameter turned out to be the value of the pressure reduction in the reduction valve. The gear ratio values, and the size of the buffer tank were also important, but to a lesser extent.

In the fast-driving variant, in addition to the amount of compressed air, the size of the buffer tank turned out to be of key importance. Gear shift time and the value of earlier opening of the drain and inlet valves had a lower influence on travel time.

It is worth mentioning that the simulation shows that there is no need to use a variator, because it is relatively easy to control the speed of the vehicle through the parameters of the pneumatic system, and the actuator can generate high torque on the drive wheel immediately in the run-up phase of the vehicle.

The simulations show a trend towards getting the longest range of the vehicle and the highest speed of the vehicle. The results are going to be compared with the experimental data while the vehicle is running.

References

[1] Fang Y, Lu Y, Roskilly A P and Yu X 2021 A review of compressed air energy systems in vehicle transport *Energy Strategy Reviews* **33** 100583

[2] Wasbari F Bakar R A, Gan L M, Tahir M M and Yusof A A 2017 A review of compressed-air hybrid technology in vehicle system *Renewable and Sustainable Energy Reviews* **67** p 935-953

[3] Chenab H, Dingac Y, Lia Y, Zhangab X and Tanb C 2011 Air fuelled zero emission road transportation: A comparative study Applied Energy **88** p 337-342

[4] Nehler T Linking energy efficiency measures in industrial compressed air systemswith non-energy benefits–A review 2018 *Renewable and Sustainable Energy Reviews* **89** p 72-97

[5] Yusop M 2006 Energy Saving for Pneumatic Actuation using Dynamic Model Prediction (Cadriff Univesity, Wales, UK) p 1-9

[6] Boes M K, Islam M, Li Y D and Hsiao-Wecksler E T 2013 Fuel Efficiency of a Portable Powered Ankle-Foot Orthosis *2013 IEEE 13th International Conference on Rehabilitation Robotics* (Seattle)

[7] Cummins J J, Nash C J, Thomas S, Justice A, Mahadevan S, Adams D E and Barth E J Energy conservation in industrial pneumatics: A state model for predicting energetic savings using a novel pneumatic strain energy accumulator *Applied Energy* **198** p 239-249

[8] Yang A, Pu J, Wong C B and Moore P 2009 By-pass valve control to improve energy efficiency of pneumatic drive system Control Engineering Practice **17** p 623-628

[9] Szakacs T 2019 Modelling and Validation of a Pneumobil *Proceedings of 3nd Agria Conference on Innovative Pneumatic Vehicles – ACIPV 2019* ed László Pokorádi (Eger) p 31-35

Application of the theory of stochastic processes and Monte-Carlo simulations for the analysis of the operation of charging stations for electric vehicles

László HANKA

Obuda University, Banki Donat Faculty of Mechanical and Safety Engineering, Institute of Mechatronics
The University of Public Service, Faculty of Military Science and Officer Training, Department of Natural Science

hanka.laszlo@uni-obuda.hu, hanka.laszlo@uni-nke.hu

Abstract. Considering the development of the Electric Vehicle industry, and the growing number of such vehicles in public transport, the charging problem of these vehicles is essential. The service process of a charging station seems a fundamental question, that must be analyzed.

Using the theory of stochastic processes, especially the Poisson process and the queuing theory; the characterization of the process can be studied. Some fundamental system parameters like the average waiting time, the average time in the system, the average time in the queue, and the average length of the queue, etc., can be modeled by the queuing theory using the service rate, the charging rate, and the number of servers, etc., as parameters.

Using the conclusions of this model, and considering the demands of Electric Vehicle users, the necessary/minimal/optimal number of servers, parking lots, etc. can be estimated. Using the theory of the Markov process, a stochastic model can be proposed for studying the long-term characterization of Electric Vehicle owners' charging behavior; that which's problem is strongly affected by the properties of charging stations. In this article, the queuing model and the Monte-Carlo method are presented for characterizing a charging station. For every calculation, Matlab has been used.

1. Introduction

Due to the growing number of electric vehicles, the charging of these vehicles is a fundamental problem, that efficiently must be solved. For example, in Hungary, the number of electric vehicles (pure electric and plug-in hybrid) in 2021 was around 30 thousand [1]. Comparing this quantity and the total number of vehicles in Hungary, which is around 4 million [2], the ratio of electric vehicles, which is around 0,75% seems negligible. According to worldwide statistics [3], recently the global ratio of electric car sales is 3%, which will reach 10% by 2025 and predicted to rise up to 58% by 2040. Due to trends for recent years, the profile change of car factories and the increasing number of such vehicles, the charging of these cars will be an essential problem of public transport soon [4]. This is a new challenge for existing, already working and, potential service providers. New charging stations must be constructed. But there are some expectations for these charging stations. On the one hand, which is fundamental for the provider, it must be economical and profitable. On the other hand, from the customers' point of view, it must be efficient, as fast as possible, there must be enough parking slots, and the waiting time must be as short as possible. In this article, the characterization of charging stations are given. During this investigation, various system parameters are considered. System parameters are properties of the station that essentially affect the operation of the station. The following parameters have been selected: the number of servers, in other words, the number of charging plugs, the number of parking lots and the charging time. The charging time is partially determined by the type of charging device (several fast-charging types exist), therefore it can be considered as a property of the station.

The paper is organized as follows: Section 2 presents the application of the basic mathematical model, the part of the mass service theory, the queuing theory [5, 6, 7], which is a specific stochastic process [8, 9], a continuous-time Markov process. In section 3, using the theory described in section 2, the effect of system parameters for the key features of the operation of a charging station will be studied. These key features, which are basic for customers, are as follows: the waiting time, the length of the queue, and the occurrence of the event when the station is full, therefore the arriving vehicle must be rejected. In section 4 the Monte-Carlo simulation method is applied [10], which provides more information about a charging station, it characterizes the charging station from another point of view. Section 5 summarizes conclusions and sketches the possible future work.

2. The *M*/*M*/*s*/*K* finite queue model

For modeling a charging station, an extremely suitable mathematical tool is the queuing theory [5, 6]. Customers (vehicle owners) sit in a queue and wait for the service. This process is depicted in Figure 1., assuming that the number of servers (charging plugs) is s, and the number of parking slots is K including places where cars are being charged.

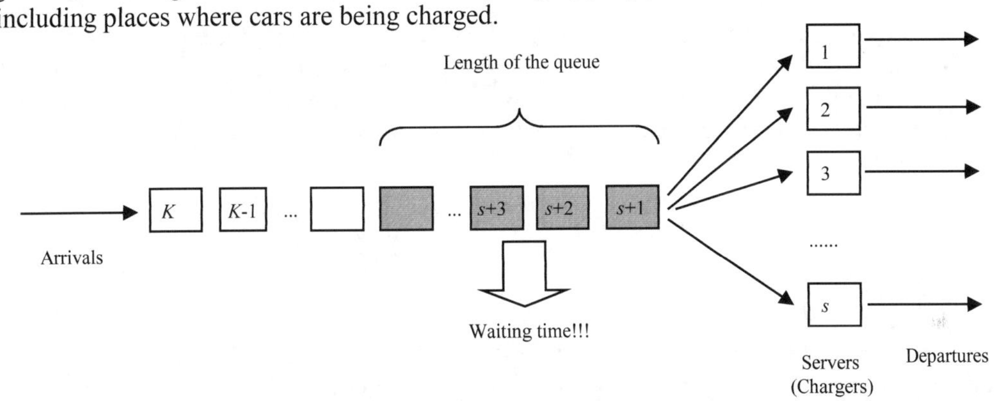

Figure 1. The illustration of the charging station as a multiple server queuing system

In Figure 1., the structure of a queuing system can be seen, including the queue itself at the station, servers, parking slots, etc. A gray rectangle refers to, that the parking slot is busy. White means that the parking slot is free. For a customer, only the number of gray slots matter, because it is the length of the queue, and the waiting time which basically depends on it. Clearly, if there are free servers, there is no queue, no waiting time, so only the illustrated situation is interesting in practice.

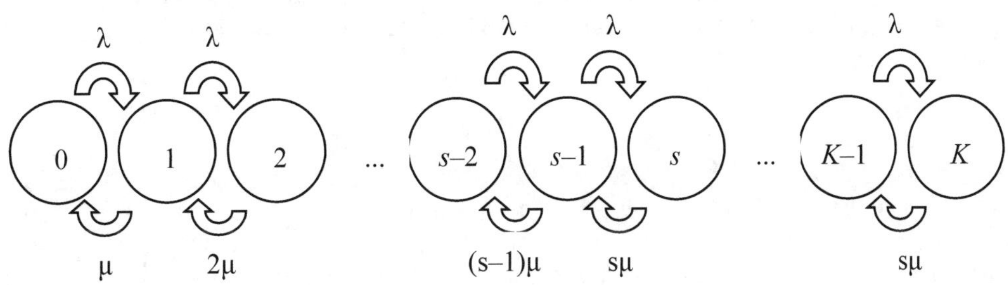

Figure 2. The rate diagram of death and birth process for the *M*/*M*/*s*/*K* model

IOP Conf. Series: Materials Science and Engineering 1237 (2022) 012002 doi:10.1088/1757-899X/1237/1/012002

The basis of this theory is the "birth and death process". The meaning of "birth" is a new vehicle in the system (arrival), and the "death" means a served vehicle (departure). This is a stochastic process, that can be illustrated by the "rate diagram" (Figure 2.). In this article the *M*/*M*/*s*/*K* queuing model is applied, as it is illustrated in Figure 1. and Figure 2. The "rate diagram" illustrates the possible states of the system, and "transition rates". The meaning of the letters and concepts are the following:

The first *M* refers to that the arrival process is "Markovian", which means that the arrival process is modeled by a Poisson distribution [11, 12], with parameter λ. The parameter λ is the "arrival rate," the average number of arrived vehicles per unit time. Clearly, the arrival rate λ is independent of the state of the system, therefore it is constant. The Poisson distribution is essential in this study, the distribution can be given by the following formula:

$$p_k = \frac{\lambda^k}{k!} e^{-\lambda}; \; k = 0,1,2,... \tag{1}$$

A basic property of a Poisson distribution is that the modeled events are "rare," which means that by selecting a short time interval, the occurrence of more than one event in the given short time interval is negligible, its probability is zero. Therefore, at any time, at most one arrival must be considered.

The basic law in probability theory [11, 12], is that if the arriving process follows a Poisson distribution, then the interarrival time - the time between two consecutive arrivals - follows an exponential distribution. The exponential distribution is also fundamental in this study, so the basic formulas are recalled. The probability density function (*pdf*) and the cumulative distribution function (*cdf*) are respectively:

$$f(x) = \begin{cases} 0; \; if \;\; x \le 0 \\ \lambda e^{-\lambda x}; \; if \; x > 0 \end{cases}; \quad F(x) = \begin{cases} 0; \; if \;\; x \le 0 \\ 1 - e^{-\lambda x}; \; if \; x > 0 \end{cases}; \tag{2}$$

The second *M* also refers to that the serving process is also "Markovian", which means that the service time is modeled by Exponential distribution, with parameter μ. The parameter μ is the "service rate", the average number of served vehicles per unit of time. Assuming that every charging device is equipped with the same type of charging plug, μ is naturally the same constant for every server. This is an essential assumption of the theory. Over the rate diagram, the number of free and busy servers must be considered, therefore the "actual" service rate depends on the current state of the system and on the current number of free servers.

The "Markovian property" is equivalent to the application of exponential distribution for the arrival and service process. A fundamental feature of the exponential distribution is the "memoryless" property. Using the language of the theory-of-probability-theory, a stochastic process is said to be Markovian = memoryless, if the following holds:

$$P\left(W > t + t' \middle| W > t'\right) = P\left(W > t\right). \tag{3}$$

This means that the probability is independent of time delay. More precisely, if *W* is the waiting time, the probability of the event, that someone has to wait longer for the service than time *t*, is the same *t'* time later. In other words, assuming that someone has already waited *t'* time for the service, the probability of the event that he/she has to wait more *t* time, is the same, independent to time *t'*. It doesn't matter when someone stands in the queue, in the morning, afternoon or tomorrow, the probability for this characteristic of the station, is the same.

This is a common property of exponential distribution and the continuous-time Markov process. This assumption is clearly valid for a queuing system. The queuing system is a continuous-time Markov process, because a new arrival or a new departure may occur at any time. A continuous-time Markov process can be defined by its "infinitesimal generator" matrix **Q**, which contains transition rates [5, 7, 9]. In the studied case, considering Figure 2., it is the following $(K+1)\times(K+1)$ quadratic matrix:

$$\mathbf{Q}=\begin{bmatrix} -\lambda & \lambda & 0 & 0 & \dots & & & & & & \\ \mu & -(\lambda+\mu) & \lambda & 0 & \dots & & & & & & \\ 0 & 2\mu & -(\lambda+2\mu) & \lambda & \dots & & & & & & \\ & & & & \dots & & & & & & \\ & & & & \dots & (s-1)\mu & -(\lambda+(s-1)\mu) & \lambda & 0 & 0 & 0 \\ & & & & \dots & 0 & s\mu & -(\lambda+s\mu) & \lambda & 0 & 0 \\ & & & & \dots & \dots & \dots & \dots & \dots & \dots & \dots \\ & & & & \dots & \dots & \dots & 0 & s\mu & -(\lambda+s\mu) & \lambda \end{bmatrix} \tag{4}$$

Since formulation of probability distributions is very complicated in general, the steady-state of the system is studied instead. The steady-state of the continuous Markov process can be given by "balance equations". These equations can be combined in an equation system that is simply the following:

$$\boldsymbol{\pi}\mathbf{Q}=\mathbf{0}; \tag{5}$$

In this system **0** is the constant zero row vector, and

$$\boldsymbol{\pi}=\left[P_0,P_1,P_2,P_3,\dots,P_K\right]; \tag{6}$$

is the probability distribution of the states of the queuing system. In this distribution, P_n is the probability of the event, that there are exactly n vehicles in the system (n = 0,1, 2, ..., K). For characterizing the charging station, the probability distribution P_n must be given. Considering (5), balance equations can be given as follows [4, 5, 6]:

$$\begin{aligned} \mu P_1 &= \lambda P_0 \\ \lambda P_0 + 2\mu P_2 &= (\lambda+\mu)P_1 \\ \lambda P_1 + 3\mu P_3 &= (\lambda+2\mu)P_2 \\ &\dots \\ \lambda P_{K-2} + s\mu P_K &= (\lambda+s\mu)P_{K-1} \\ s\mu P_K &= \lambda P_{K-1} \end{aligned} \tag{7}$$

This equation system provides the following results: Every probability can be given by the product

$$P_n = C_n \cdot P_0; \tag{8}$$

where on the one hand

$$C_n = \begin{cases} \frac{1}{n!}\left(\frac{\lambda}{\mu}\right)^n; & n = 0,1,2,...,s; \\ \frac{1}{s!s^{n-s}}\left(\frac{\lambda}{\mu}\right)^n; & n = s+1,...,K; \end{cases} \tag{9}$$

and on the other hand

$$P_0 = \left(\sum_{n=0}^{s}\frac{1}{n!}\left(\frac{\lambda}{\mu}\right)^n + \sum_{n=s+1}^{K}\frac{1}{s!s^{n-s}}\left(\frac{\lambda}{\mu}\right)^n\right)^{-1}; \tag{10}$$

The letter s, as it is already defined, refers to the number of servers of the system. One "server" means one charging plug in the charging system, therefore s is equal to the number of charging plugs at one charging station. It can be assumed in practice, that the number of servers is greater than one, therefore here a "multiple server system" is being studied and applied.

Finally, K means that the proposed model is the "finite queue" model. K is equal to the number of parking slots in the charging system, including places where vehicles are being charged. In other words, K is the maximum number of vehicles, the capacity of the charging system. Clearly, every charging system has a finite number of places for vehicles, therefore if the system is full, the next arriving vehicle can't stand into the queue, it must be rejected. Rejection of a vehicle is a key problem for the customer and the provider too. This problem will also be studied in section 3.

The finite queue model has a great advantage over the infinite queue model. For describing this advantage, the "utilization factor" must be defined. If the average interarrival time is ta, and the average service time is ts, then clearly

$$\lambda = \frac{1}{ta}; \; \mu = \frac{1}{ts}; \tag{11}$$

The utilization factor of a charging system is defined by the ratio

$$\rho = \frac{\lambda}{s\mu} = \frac{ts}{s \cdot ta}; \tag{12}$$

The utilization factor [5] is a fundamental quantity for management because it is a measure of the efficiency and effectiveness of the charging station, therefore it is desirable to be as big as possible. So, in practice, this factor should be greater than or equal to one, $\rho \geq 1$ can be considered as an everyday situation. (For example, if the average interarrival time is 10 minutes, the average service time is 60 minutes, and there are 4 charging plugs in the system, then $\rho = 1{,}5$; etc.)

The finite queue model has a great advantage against the infinite model. The infinite model makes sense only if $\rho < 1$, because if $\rho \geq 1$, according to the behavior of the geometric series, the "system blows up," the number of customers in the queuing system approaches infinity as time goes on and tends towards infinity. But the finite queue model is applicable also for the case when $\rho \geq 1$ [5, 6, 7]. This advantage will be strongly exploited in this study.

Naturally, the following questions arise: What are the key features of a queuing system, and what are the most important parameters of the charging station, which affect key features?

On the one hand, as it has already mentioned in the introduction; from the point of view of the owner of a vehicle, clearly the most essential characteristics are the waiting time and the length of the queue. Furthermore, the case when a system is full, therefore the vehicle must be rejected is also important.

On the other hand, for the service provider, fundamental parameters of the system are the number of servers (number of charging plugs = s), the number of parking slots (capacity = K), and in some sense the charging time, because if the system was equipped with fast charging devices, it would be much more efficient.

The time, spent in the system (W, including service time) and the number of vehicles in the system (L, including vehicles that are being served) are also random variables. The formulation of these random variables by their density or distribution function can be very complicated in general. This is the reason why the expected value, in other words, the mean of these random variables is studied instead of their distribution function. These mean values come from balance equations, using the classic definition of expected value in the probability theory.

It is self-evident, that for a customer (owner of a vehicle) the number of vehicles being served is not important, only the length of the "actual" queue matters. Similarly, the time spent with charging his/her vehicle is personal, depending on the owner's intention, the type of the vehicle, the charge level of the accumulator (battery), etc., so it is not a feature of the system. Therefore, only the “actual" waiting time must be investigated. The following notations and definitions will be used [5]:

- L_q: The average (mean) length of the queue in the system (excluding vehicles that are being charged).
- W_q: The average (mean) waiting time in the system (excluding service time).

These quantities can be derived from formulas (8)-(10). Using the classic definition of expected value [11, 12] of a discrete random variable. For the average (mean) length of the queue, the following is obtained:

$$L_q = \sum_{n=0}^{K} (n-s) P_n = P_0 \frac{\rho}{s!(1-\rho)^2} \left(\frac{\lambda}{\mu}\right)^s \left(1-\rho^{K-s} - (K-s)\rho^{K-s}(1-\rho)\right); \tag{13}$$

For giving the average waiting time, Little's formula [4, 5, 6] is necessary:

$$W_q = \frac{L_q}{\bar{\lambda}}; \tag{14}$$

Where $\bar{\lambda}$ is the average arrival rate in the $M/M/s/K$ system:

$$\bar{\lambda} = \sum_{n=0}^{\infty} \lambda P_n = \sum_{n=0}^{K-1} \lambda P_n = \lambda(1-P_K) \tag{15}$$

Summarizing these results the average waiting time in the queue can be given by the following expression:

$$W_q = \frac{L_q}{\lambda(1-P_K)}; \tag{16}$$

In this formula probability P_K - the probability of the event that the system is full, there are K vehicles in the system - also comes from expressions (8)-(10).

3. Characterization of a charging station

In this section, the *M/M/s/K* queuing system will be applied as a model for characterizing the charging station. The basis for every further calculation is presented in the previous section by formulas (1)-(16). Calculations are performed by Matlab. The mean values L_q, W_q, and the probability P_K will be studied, because these are the key features of a charging station, in other words, the most important questions for a customer.

The following questions will be investigated: How these key features are affected by the essential parameters of a charging system, such as the number of servers (*s*), the charging (service) time (*ts*), and the number of parking slots (*K*). In every subsection, the connection between key features and one of these essential parameters will be studied and illustrated, using Matlab.

3.1. The effect of the number of servers

The number of servers (charging plugs) seems to be the most essential parameter of a charging station. In this section, the effect of the number of servers is studied. First, the effect on the average waiting time.

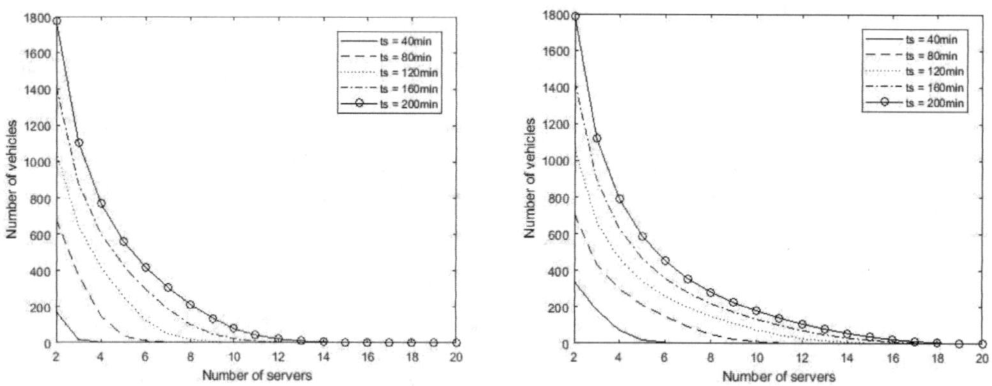

a) *ta* = 20 min, *K* = 20; **b)** *ta* = 10 min, *K* = 20;

Figure 3. The average waiting time (*Wq*) as a function of the number of servers (*s*)

Figure 3. depicts the average waiting time (*Wq*) as a function of the number of servers (*s*) for different numbers of service time (*ts*), if the capacity of the charging station (*K*) and the interarrival time (*ta*) is fixed.

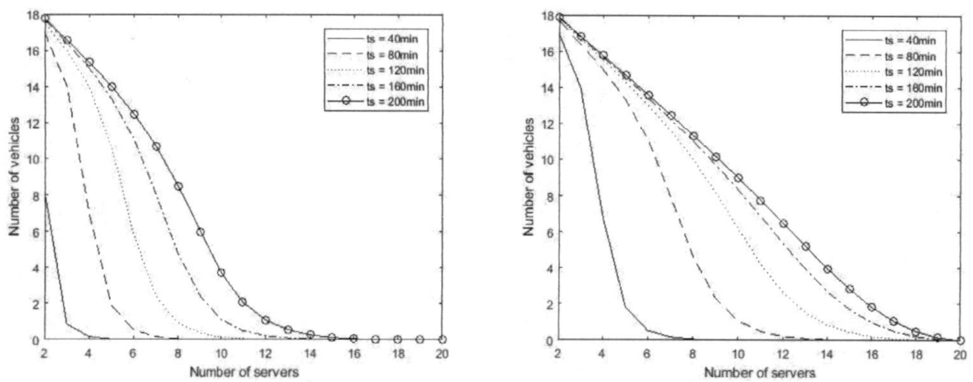

a) *ta* = 20 min, *K* = 20; **b)** *ta* = 10 min, *K* = 20;

Figure 4. The average length of the queue (*Lq*) as a function of the number of servers (*s*)

The following key feature is the length of the queue. Figure 4. illustrates the average length of the queue (*Lq*) as a function of servers (*s*), for several numbers of charging time (*ts*) but also for fixed capacity (*K*) and for fixed interarrival time (*ta*).

Finally, the probability of the event, that one vehicle must be rejected (P_K) because the charging station is full, is studied. Figure 5. depicts the probability P_K as a function of *s* for different numbers of service time (*ts*) and also for a fixed capacity (*K*) and for fixed interarrival time (*ta*).

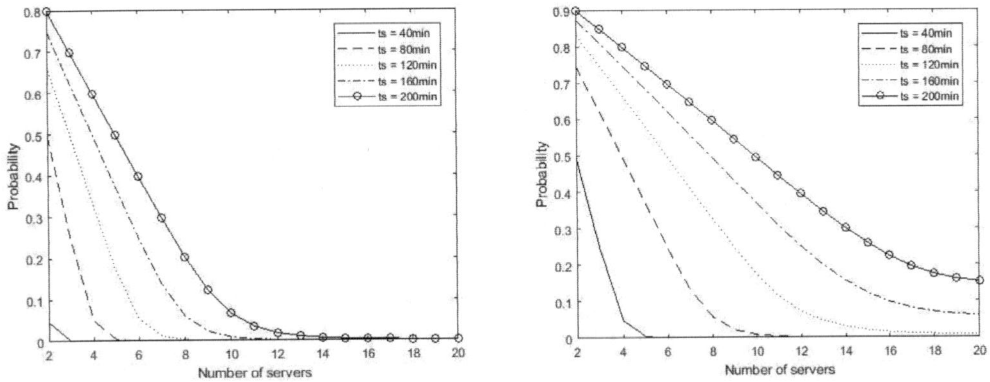

a) *ta* = 20 min, *K* = 20; **b)** *ta* = 10 min, *K* = 20;

Figure 5. The probability of rejection of a vehicle (P_K) as a function of the number of servers (*s*)

It can be seen on these figures, that either the average waiting time or the average length of the queue and also the probability of the rejection of a vehicle is a decreasing function of the number of servers. These features are significantly affected by the number of servers, especially if *s* is a small value. But clearly what is "small" depends on the average interarrival time. The situation is worse if the arrival rate is greater. These relationships can be used for estimating the necessary/desirable/"optimal" number of servers at a station.

3.2. The effect of the charging (service) time (ts)

The length of the charging time is also a fundamental parameter of a charging station, but obviously depends not only on the type of chargers, but also on the intentions of customers, types of vehicles, and the charge level of the accumulators (batteries), which are random. Here the role of the first aspect is emphasized. In this section, the effect of the charging time is studied, first of all, the effect on the average waiting time.

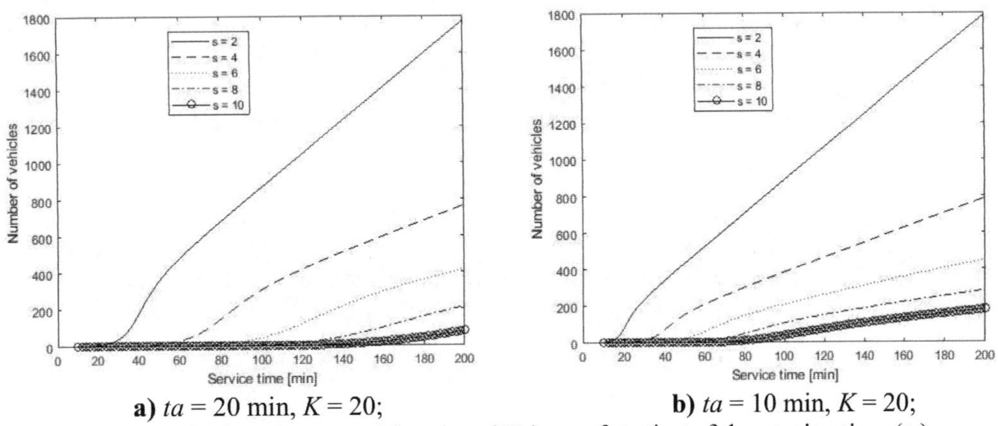

a) *ta* = 20 min, *K* = 20; **b)** *ta* = 10 min, *K* = 20;

Figure 6. The average waiting time (*Wq*) as a function of the service time (*ts*)

Figure 6. depicts the average waiting time (*Wq*) as a function of the service time (*ts*) for different numbers of servers (*s*), if the capacity of the charging station (*K*) and the interarrival time (*ta*) is fixed.

The following key feature is the length of the queue. Figure 7. illustrates the average length of the queue (*Lq*) as a function of the service time (*ts*), for several numbers of servers (*s*) but also for fixed capacity (*K*) and for fixed interarrival time (*ta*).

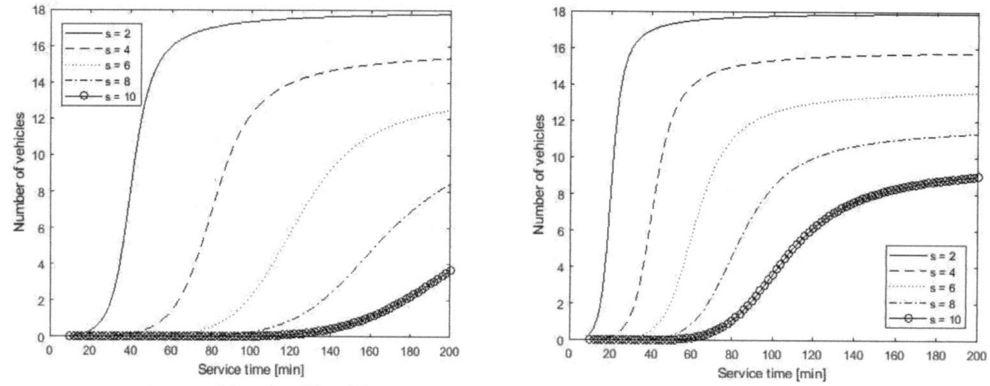

a) *ta* = 20 min, *K* = 20; **b)** *ta* = 10 min, *K* = 20;

Figure 7. The average length of the queue (*Lq*) as a function of the service time (*ts*)

Finally, the probability of the event, that a vehicle must be rejected (P_K) because the charging station is full, is studied. Figure 8. depicts the probability P_K as a function of *ts* for different numbers of servers (*s*) and also for a fixed capacity (*K*) and for fixed interarrival time (*ta*).

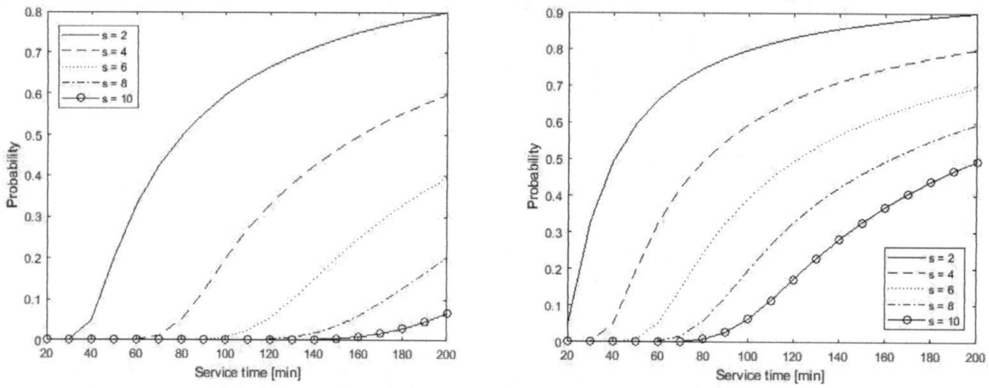

a) *ta* = 20 min, *K* = 20; **b)** *ta* = 10 min, *K* = 20;

Figure 8. The probability of rejection of a vehicle (P_K) as a function of the service time (*ts*)

It can be seen on these plots, that either the average waiting time or the average length of the queue and the probability of the rejection of a vehicle is an increasing function of the service time. These features are significantly affected by the length of the service time, especially if *ts* exceeds a specific value that depends on the number of servers. The situation is again worse if the arrival rate is greater. These relationships can be used for selecting the appropriate/suitable type of charger.

3.3. *The effect of the number of parking slots*

The effect of the number of parking slots, the capacity of the charging station is the last studied quantity that is also considered as a fundamental system parameter. In practice at an ordinary charging station, there are more parking slots, than the number of servers, providing parking possibilities for those who wait for the service. In this section, the effect of this parameter is being considered.

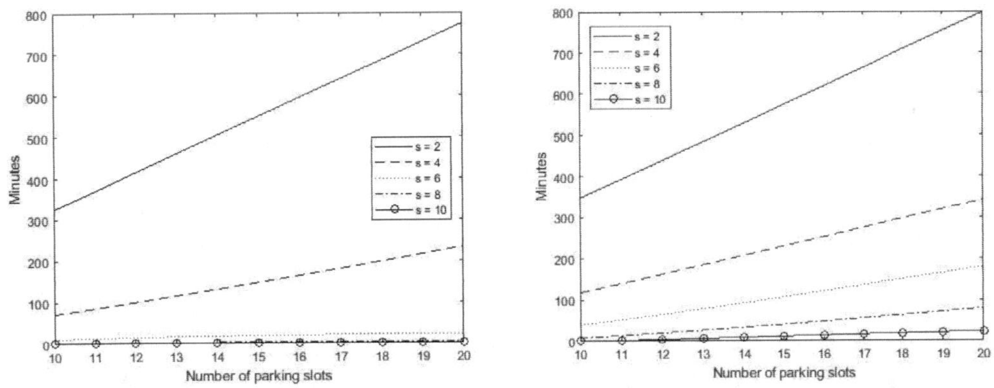

a) *ta* = 20 min, *ts* = 90 min; **b)** *ta* = 10 min, *ts* = 90 min;

Figure 9. The average waiting time (*Wq*) as a function of the capacity (*K*)

Figure 9. depicts the average waiting time (*Wq*) as a function of the capacity (*K*) for different numbers of servers (*s*) if the service time (*ts*) and the interarrival time (*ta*) is fixed.

The following key feature is the length of the queue. Figure 10. reflects the average length of the queue (*Lq*) as a function of the capacity (*K*), for several numbers of servers (*s*) but also for fixed service time (*ts*) and for fixed interarrival time (*ta*).

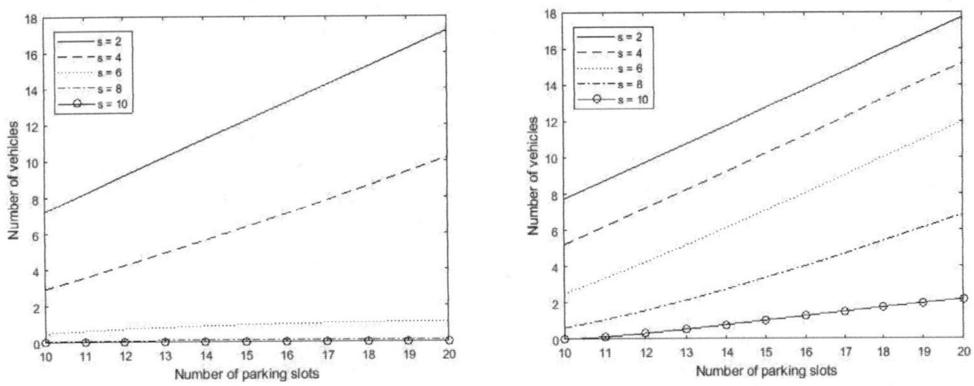

a) *ta* = 20 min, *ts* = 90 min; **b)** *ta* = 10 min, *ts* = 90 min;

Figure 10. The average length of the queue (*Lq*) as a function of the capacity (*K*)

It can be seen on these plots, that either the average waiting time or the average length of the queue is an increasing function of the capacity, if the arrival rates and the service times are fixed. These features are also affected by the number of parking slots. But, comparing this case and the previous subsections, the nature of the effect is different. In this case, the function is close to linear. The situation is again worse if the arrival rate is greater. These relationships can be used for planning the appropriate/necessary number of parking slots.

Finally, the probability of the event, that a vehicle must be rejected (P_K) because the charging station is full, is studied. Figure 11. depicts the probability P_K as a function of K for different numbers of service time (ts) and a fixed number of servers (s) and for a fixed interarrival time (ta).

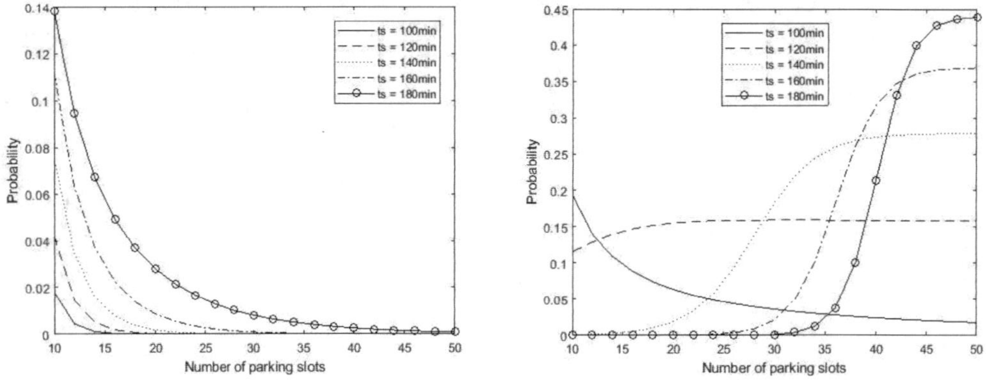

a) $ta = 20$ min, $s = 10$; **b)** $ta = 10$ min, $s = 10$;

Figure 11. The probability of rejection of a vehicle (P_K) as a function of the capacity (K)

Considering Figure 11. it can be visualized that the behavior of the probability-capacity function strongly depends on the arrival rate. The probability of rejection via capacity is more complicated than previous connections. Its monotonicity depends on the arrival rate, as it can be seen in Figure 11. for a fixed number of servers. So, if someone plans a charging station, the necessary number of parking slots seem to be a critical problem, which depends on the arrival rate.

3.4. The effect of the utilization factor

The utilization factor (ρ) measures the efficiency of the charging station, ("to what extent it is utilized"), which is a basic economic property for the providers. The greater the utilization factor the more efficient the station is; therefore, it must be as big as possible. According to (12) ρ depends on the average interarrival time, the average service time, and the number of servers.

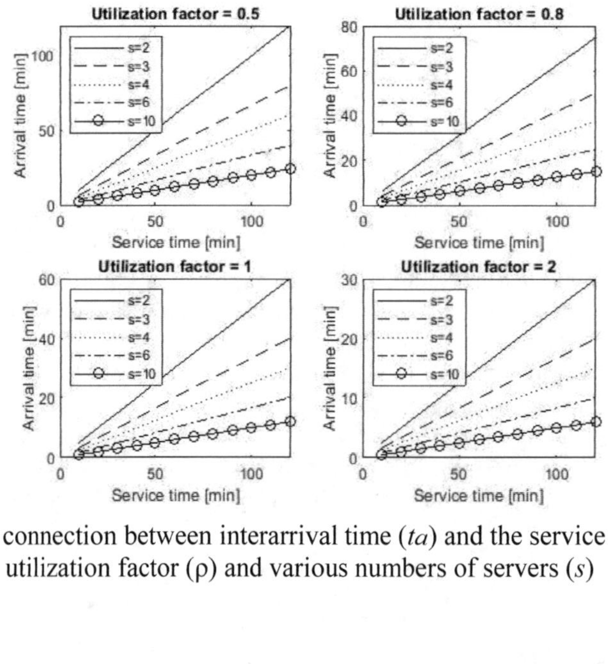

Figure 12. The connection between interarrival time (ta) and the service time (ts) for fixed utilization factor (ρ) and various numbers of servers (s)

Figure 12. depicts the corresponding *ta* and *ts* values for several numbers of servers and some fixed ρ values. These plots facilitate visualizing some cases. The point is that one key feature, the average length of the queue (*Lq*), can be considered as a function of the utilization factor (ρ). In other words, this feature depends explicitly on ρ [5]. This is not true for the average waiting time! Figure 13. visualizes the average length of the queue as a function of the utilization factor.

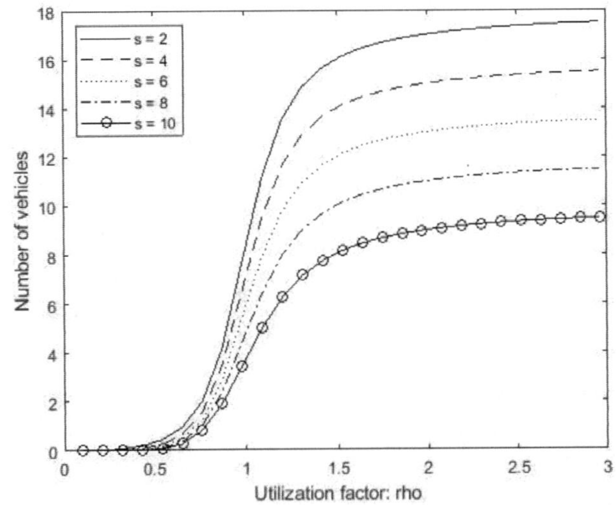

Figure 13. The average length of the queue (*Lq*) as a function of the utilization factor (ρ)

Considering Figure 13. it is obvious, that the average length of the queue is significantly affected by the utilization factor, and surprisingly, the sharply increasing section of the function is in the neighborhood of ρ = 1, independently to the number of servers. Beyond ρ = 1,5 the function becomes nearly constant for every *s* value.

4. Monte-Carlo simulations

In section 3. the average/mean value of several probability distributions has been studied because the formulation of the distribution functions can be very complicated. The expected value is essential information for a probability distribution, but naturally, the distribution itself provides more information about the random variable than the mean, which is only one single real number. In this section, instead of giving complicated formulas for distributions, the Monte-Carlo simulation method is applied for investigating some previously applied distributions. The basis of the Monte-Carlo method is the simulation of random numbers [10]. Obviously, an accurate distribution is not equivalent to a random simulation. But, comparing the Monte-Carlo method [12, 13] to the "mean value approach," the Monte-Carlo method provides more information about random variables for practice from another perspective, which can be interesting for the service provider and for customers too.

4.1. Simulation of the waiting time

In section 3. the average waiting time has been investigated. According to the theory, the conditional waiting time in the queue, assuming that the waiting time is positive, is a random variable that follows exponential distribution [5]. If the parameter of the exponential distribution is known, which is the reciprocal of the expected value, in other words, the average waiting time; then the distribution can be simulated using the Monte-Carlo method by the following way.

The cumulative distribution function is given by (2). If a uniformly distributed random number r is generated in the interval [0, 1[, then the inverse of the equation (2), which is

$$r = 1 - e^{-\lambda W} \Rightarrow W = \frac{\ln(1-r)}{-\lambda} \tag{17}$$

Which provides an exponentially distributed random variable, where W is the random waiting time. So, if $r \in [0, 1[$, then W provided by (17) follows an exponential distribution. It must be emphasized that in this formula λ is a general notation for the parameter of exponential distribution and not the arrival rate!!! If the expected value as the mean waiting time is given, then λ is it's reciprocal.

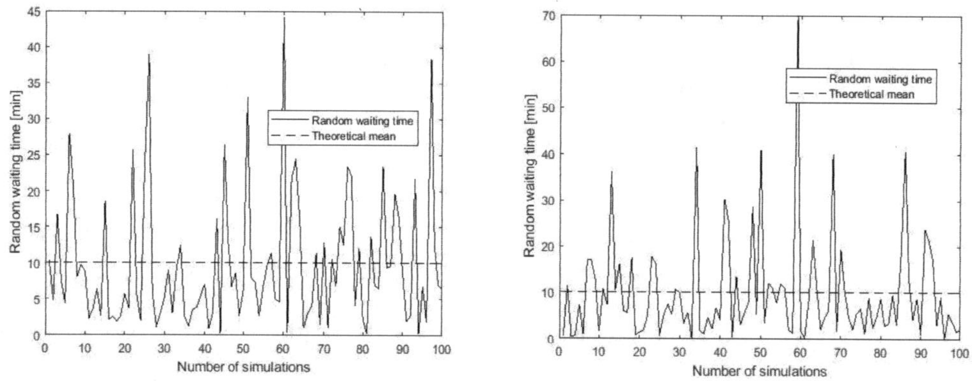

Figure 14. Exponentially distributed random waiting time, if $\lambda = 1/10$ [1/min]

Figure 14. depicts two sequences of experiments. Both plots illustrate 100 random trials on the basis of (17), assuming that, for example, the average waiting time is 10 minutes. The only difference between these plots is that they are two independent sequences of experiments, but the parameter is the same. The horizontal dashed line illustrates the average waiting time. The point is, which is not self-evident for those who don't deal with probability theory, is that the actual waiting time can be greater, even much greater than the average value; considering these simulations, the number of events when the actual waiting time is longer than the average is 33 and 34, and the longest waiting time is 45 and 70 minutes respectively.

The first result is "natural", because the accurate probability of the event, that the waiting time is longer than the average is the following:

$$P\left(W > \frac{1}{\lambda}\right) = 1 - F\left(\frac{1}{\lambda}\right) = 1 - \left(1 - e^{-\lambda \frac{1}{\lambda}}\right) = e^{-1} \approx 0{,}3678 \tag{18}$$

Multiplying this probability by the number of trials, the result is absolutely consistent with the theory. But the second result, that the real waiting time can be up to 5 or 6 times the average waiting time, or even more, is also a remarkable property of the exponential distribution. Consequently, it must be underlined, that the average waiting time as an expected value, provides a fundamental characterization of the random variable. But, for customers in a queuing system it should be clear, that the actual waiting time can be much greater sometimes, and the factor can be up to 6 or even higher.

IOP Conf. Series: Materials Science and Engineering 1237 (2022) 012002 doi:10.1088/1757-899X/1237/1/012002

4.2. Simulation of the total number of vehicles in the charging system

In subsection 4.1. a simple exponential distribution has been simulated by random numbers, using the formula (17). This simple method can be used as a basis for studying a more complex system, the charging station itself by random numbers. Assuming that the number of arrivals follow a Poisson distribution, the interarrival time follows an exponential distribution, with parameter λ (arrival rate), and that the service time also follows an exponential distribution, with parameter μ (service rate), for every server; the evolution of the number of vehicles can be simulated by the Monte-Carlo method.

The applied procedure is the following: roughly speaking, $s + 1$ exponential distributions must be simulated and must be combined. More precisely, considering λ (arrival rate) for arrivals, and taking into account **every server**, considering μ (service rate) for the departures, using formulas

$$ta = \frac{\ln(1-r)}{-\lambda}; \quad ts_k = \frac{\ln(1-r)}{-\mu}; \quad k = 1,2,3,\ldots,s; \tag{19}$$

arrivals and departures are simulated for every server at the same time, simultaneously. If one vehicle arrives, a new interarrival time is simulated when the following vehicle arrives, and so on. Similarly, if one vehicle is served and departs, a new service time is simulated, when a new vehicle departs, etc., for every server independently and simultaneously. If a new vehicle arrives or a vehicle is served, the total number of cars are calculated, and if there are free servers, the simulation of service times for those servers is suspended, etc. This is how the evolution of the total number of vehicles at the charging station can be simulated by the Monte-Carlo method.

For illustration three cases have been selected. The evolution of the number of the vehicles is illustrated if $\rho < 1$, $\rho = 1$ and if $\rho > 1$ respectively, where ρ is the utilization factor (12). As typical parameters, $s = 10$ and $ts = 120$ min have been chosen, and several interarrival times (ta). For possible further study, it is assumed that the capacity of the station is twice the number of servers.

Figure 15. illustrates the evolution of vehicles at a charging station if $ta = 15$ min, therefore if $\rho = 0{,}8 < 1$. The initial number of vehicles is $N_0 = s$. The evolution is clear, one step up means one arrival, and one step down means one served vehicle. The growth can be at most one, due to the property of the Poisson distribution [10, 11], but the decrease can be greater than one because there are s servers, as is presented in Figures 15-17.

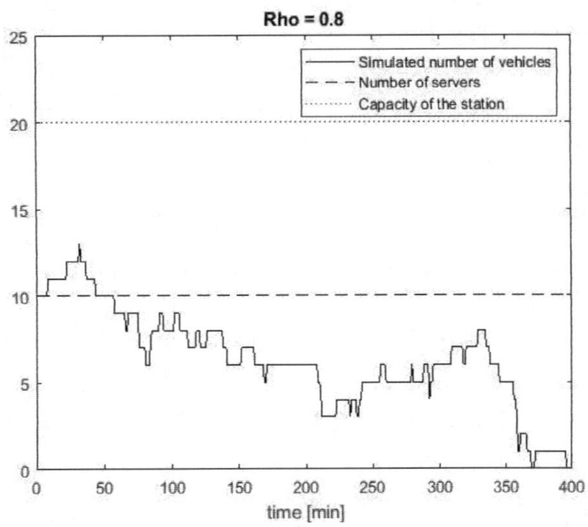

Figure 15. Evolution of the number of vehicles if $\rho = 0{,}8$ and $N_0 = s$

Figure 16. reflects the evolution of vehicles at a charging station if ta = 12 min, therefore if $\rho = 1$. The initial number of vehicles is also $N_0 = s$.

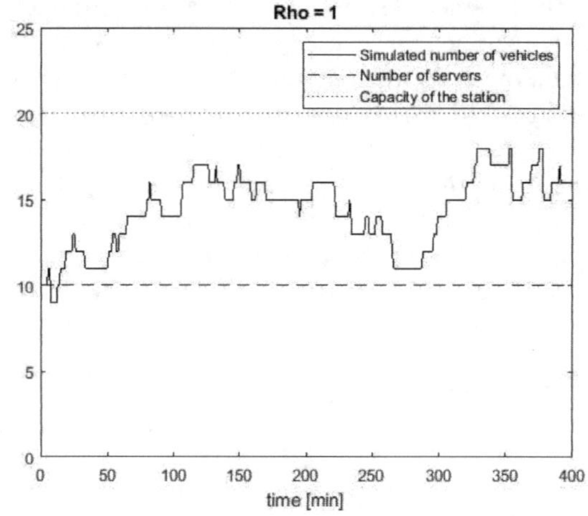

Figure 16. Evolution of the number of vehicles if $\rho = 1$ and $N_0 = s$

Finally Figure 17. illustrates the evolution of vehicles at a charging station if $ta = 6$ min, therefore if $\rho = 2 > 1$. In this case the initial number of vehicles is $N_0 = 0$.

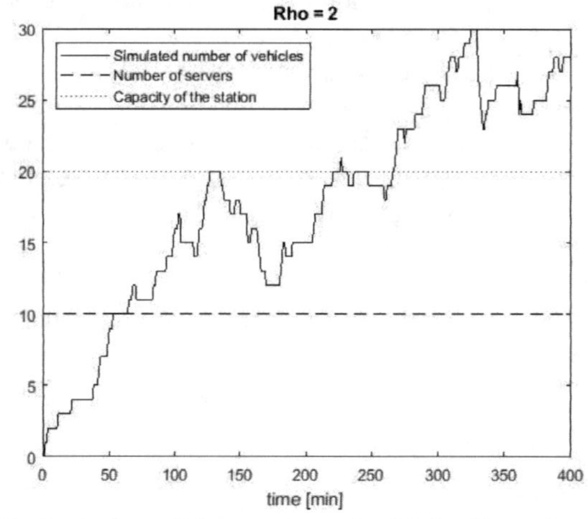

Figure 17. Evolution of the number of vehicles if $\rho = 2$ and $N_0 = 0$

These figures demonstrate typical trends and clearly describe the importance and meaning of the utilization factor. The advantage of the Monte-Carlo method is that the random simulation provides more information about the behavior of the system, the evolution of the number of vehicles at the charging station, than a simple expected value. For example, considering the assumption, that the capacity $K/s = 2$, or any other ratio, the obtained plots help study the problem of the necessary/"optimal"/desirable number of parking slots, and help reduce the probability of rejection.

5. Conclusions

In the light of what has been presented in previous sections, it is demonstrated that the queuing theory, especially the finite queue model and the Monte-Carlo simulation method are suitable and efficient tools for describing the characteristics of a charging station. Results, presented in sections 3.

and 4. help investigate the operation of existing charging stations and help plan and construct a new station, that meets the expectations and needs of customers and providers.

Considering system parameters such as the number of servers, the charging time, and the capacity of the station; it has been presented that these parameters affect the operation and efficiency of the charging station. The quantitative relationship can be estimated by the obtained graphs. The key features of the station, such as the length of the queue, the waiting time, and the probability of rejection of a vehicle are essentially affected by system parameters, which are especially true for the number of servers and the charging time. The nature of the effect of capacity is different because as it was presented, the monotonicity of the probability-capacity function is not unique, it depends on more parameters.

The obtained functions are only depicted by plots, but for users, these curves are enough for studying properties in question, and for numerical estimations.

For example, summarizing some results from section 3, it is obvious, that for improving the efficiency of a charging station an increasing the number of chargers are required, while decreasing the charging time. But, depending on the actual/general/average interarrival time, that comes from experience, there is a reasonable upper limit for the number of chargers, and a reasonable lower limit for the length of charging time, because beyond specific limits, key features don't change significantly. The decision depends on the user and the provider.

The Monte-Carlo simulation extends this picture with some more data. This procedure provides information not only about the expected values but about the behavior of random variables, such as the exponential and Poisson distribution, and what is more important; the overall behavior of the charging station, such as the evolution of the numbers of vehicles in the system. This study also helps estimate the necessary capacity of a station and reduce the probability of rejection. Furthermore, it points out the role and importance of the utilization factor.

The Author's proposed prospective future research direction is the study of applicability of discrete time Markov chains, for investigating the long-term behavior of the distribution of electric vehicles between home and public charging stations.

References

[1] Villanyautósok, 2021: https://villanyautosok.hu(2021/12/10)
[2] Központi Statisztikai Hivatal 2021: https://www.ksh.hu/stadat_files/sza/hu/sza0040.html (2021/12/10)
[3] Electric Car Statistics 2021: https://policyadvice.net/insurance/insights/electric-car-statistics (2021/12/10)
[4] Nagy I. 2021: Hierarchical Mapping of an Electric Vehicle Sensor and Control Network, *Acta Polytechnica Hungarica*, **18**. *161-180.*
[5] Hillier-Lieberman 2001 Introduction to operations research; McGraw-Hill, New York
[6] Sztrik J. 2011 Basic queueing theory; University of Debrecen
[7] Cooper 1981 Introduction to queueing theory; North Holland; New York
[8] Karlin-Taylor 1998 An introduction to stochastic modeling; Academic Press, USA
[9] Knill 2009 Probability theory and Stochastic Processes; Overseas Press, India,
[10] Rubinstein-Kroese 2017 Simulation and the Monte Carlo methods; Wiley, USA
[11] Vetier A. 2011 Probability theory with simulations. BME, Budapest
[12] Jaynes 2003 Probability theory; Cambridge UP
[13] Pokorádi L., Molnár B. 2010 A Monte-Carlo szimuláció szemléltetése. *Szolnoki Tudományos Közlemények* **14** pp. 1-12. , 12 p.
[14] Pokorádi L., Molnár B. 2011 Monte-Carlo simulation of the pipeline system to investigate water temperature's effects. *The Polytechnical University of Bucharest. Scientific Bulletin. Series D: Mechanical Engineering* **73**: 4 pp. 223-236. , 14 p.

Matrix representation of an EV's Sensory and Control System

István Nagy

Óbuda University, Bánki Donát Faculty of Mechanical and Security Engineering,
Institute of Mechatronics and Vehicle Engineering, Dept. of Mechatronics
H1081 Budapest, Népszínház street 8.;
Email: dr.nagy.istvan@uni-obuda.hu; ORCID: 0000-0002-8250-7969

Abstract. This article will be the continuation of previously published papers and journal articles. The previous developments have been finished by the suggestion of possible matrix-representation of the sensory and control block diagram system of the examined, fully electric vehicle[1] (EV) [1], [2], [3], [5]. This article will show, how these block diagrams can be represented by some special matrices. The first step in the developments was the creation of the sensory and control map of the system (in this form of expression the "system" is the electrical network of the vehicle). Later, this electrical network, based on functionality, was segmented into the different subsystems, like: EV Control System, Driver Control System, ..., see [2]. Hereupon the subsystems could be represented by the graph of the subsystem and, as it is well known, each graph can be represented as a matrix. The problem arose from the fact, that these subsystems are overlapped with each other. The present article will show, how these overlaps can be handled by a special matrix representation.

1. Introduction and Theoretical Background

At the beginning let me introduce some basic definitions regarding graph theory. In this paper the author will consider the graphs consisting of a set of nodes (vertex - v) and edges (e). Symbolically it can be written: $G=(v,e)$. Based on a created network of points (v) and lines (e) (what is basically the graph), -including, but limited-, it can be distinguished as several types of graphs, like: grid-like graph, hierarchical tree graph, polygon type graph, connected- and disconnected graphs, hypergraphs, multigraphs, ...etc. [4]. Another aspect concerning graphs is arising from orientation of edges, where we can talk about directed and undirected graphs, moreover regarding weighting of edges, we can talk about weighted and unweighted graphs.

1.1. Mathematical representation of graphs

Each finite graph can be mathematically defined by some type of *matrix* or some *list*. The most used matrices are the *adjacency* matrix and the *incidence* matrix, and for linked representation is known as an *adjacency list* [8].

1.1.1. Adjacency matrix

In the case of an adjacency matrix, the elements of the matrix are created from the pairs of vertices (v), (the nodes of the graph). That created matrix will be a square matrix. Usually the adjacency is represented by "*1*" or "*0*". Typically, the diagonal of the matrix contains zeros (if no loop on nodes). In case of an *undirected* graph, the adjacency matrix is *symmetric*. The "*0*" and "*1*" numbers, of

[1] The mentioned EV is a *NISSAN LEAF Z0,* owned by the Óbuda University.

IOP Conf. Series: Materials Science and Engineering 1237 (2022) 012003 doi:10.1088/1757-899X/1237/1/012003

elements of the matrix, come from the convention that each edge, connecting two nodes, adds "*1*" to the cell in the matrix, resp, "*0*" where there is no edge between the nodes.

In the case of a *directed* graph, the direction can be represented 2 ways, see equations (1) and (2). Mathematically it can be defined as follows: Let have a graph $G=(v_{(i)}, e_{(i)})$; Where $\boldsymbol{V}=[v_{(i)}]$ is the vector of vertices (nodes); $\boldsymbol{E}=[e_{(j)}]$ is vector of edges. Let $\boldsymbol{A}=[a_{(i,j)}]$ the adjacency matrix. There exists several definitions for matrix representation of the graphs. In case of equation (1) the matrix contains only "0"-s and "1"-s, because the "incoming" edges are "0" represented, while in equation (2) the $a_{(i,j)} = -1$, if the edge $e_{(j)}$ leaves vertex $v_{(i)}$; and $a_{(i,j)} = 1$ if it enters vertex $v_{(i)}$, otherwise the element is "0". This representation is more useful for technical applications. (Sometimes the "+" and "-" interpretations may be interchanged.)

$$a_{i,j}\begin{cases} 1 & v_i \underset{e_j}{\rightarrow} \\ 0 & otherwise \end{cases} ; \qquad (1)$$

$$a_{i,j}\begin{cases} 1 & v_i \underset{e_j}{\rightarrow} \\ -1 & v_i \underset{e_j}{\leftarrow} \\ 0 & otherwise \end{cases} ; \qquad (2)$$

In case of *weighted graphs,* instead of "*1*"-s, the weight of the appropriate edge is usually used. In practical use, the edges can represent connection wires (electrical engineering, informatics) and the weights should be expected voltage/current values or bit-strings.

1.1.2. Incidence matrix

The most important difference between the adjacency and incidence matrices is, that as long as the adjacency matrix is created from pairs *node × node*, the incidence matrix is encoding the relations between the *nodes × edges*. This will be resulting, that usually the incidence matrix is not a square matrix, rather an $n \times m$ matrix, where "*n*" represents the number of nodes, while "*m*" the number of edges. The values of the matrix elements can be, -in particular case-, "*1*", if vertex $v_{(i)}$ and edge $e_{(j)}$ are incident, otherwise "*0*". For *oriented* and *weighted* graphs to be valid, the same rules that were stated previously in case of adjacency/incidence matrices, by adding that: the orientation of the edge is signed by "+" resp. "-", depending on orientation, (in technical applications the *bi-directional* transfer will be signed by "±1") and in case of weighting, instead of "1" the weights are written also.

For a better understanding, the theory described before, will be illustrated through a short example, where a simple electric circuit will be converted into the directed graph, what will be mathematically expressed in an adjacent (**A**) and incident (**B**) matrices, using representation stated in equation (2).

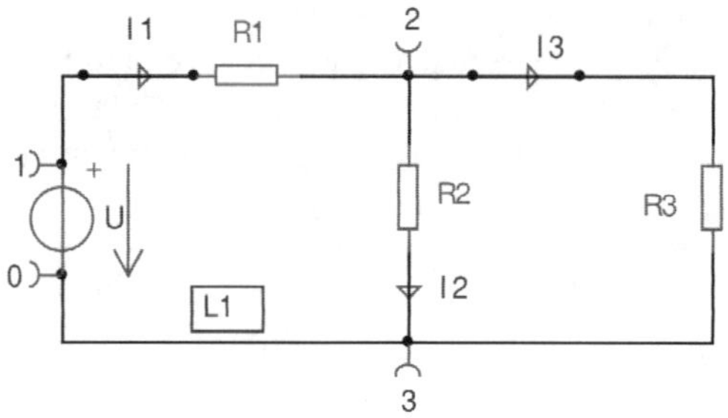

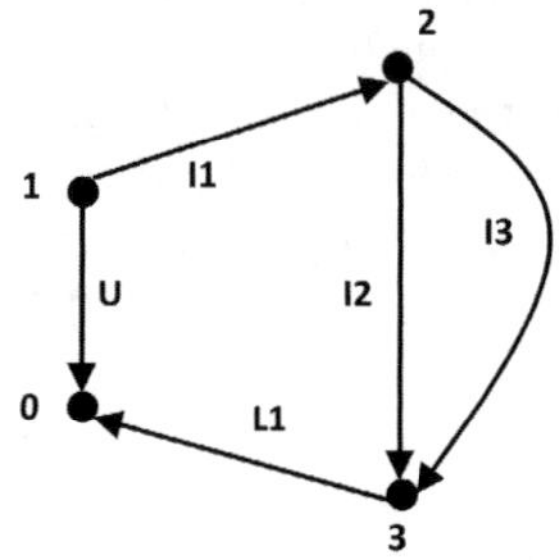

Figure 1. Electrical circuit and corresponding graph

On Fig. 1, the *nodes* are: *0, 1 ,2 ,3* and the *edges*: U, I_1, I_2, I_3, L_1. The notations are set so that the example can be used later for Kirchoff Current Law (KCL) calculations. And now the matrices: the corresponding adjacency matrix $\mathbf{A}=\{a_{(i,j)}\}$ where the rows (i) and columns (j) are created from the vertices $v_{(i,j)}$, where $i,j=\{0,1,2,3\}$.
In case of incidence matrix $\mathbf{B}=\{b_{(n,m)}\}$, the rows $n=\{0,1,2,3\}$ are created from nodes, and the columns $m=\{U, I_1, I_2, I_3, L_1\}$ are formed from the edges.

$$A_{(4x4)} = \begin{matrix} nodes \\ 0 \\ 1 \\ 2 \\ 3 \end{matrix} \begin{matrix} \begin{matrix} 0 & 1 & 2 & 3 \end{matrix} \\ \begin{bmatrix} 0 & -1 & 0 & -1 \\ 1 & 0 & 1 & 0 \\ 0 & -1 & 0 & 1 \\ 1 & 0 & -1 & 0 \end{bmatrix} \end{matrix} \qquad B_{(4x5)} = \begin{matrix} nodes/edges \\ 0 \\ 1 \\ 2 \\ 3 \end{matrix} \begin{matrix} \begin{matrix} U & I1 & I2 & I3 & L1 \end{matrix} \\ \begin{bmatrix} -1 & 0 & 0 & 0 & -1 \\ 1 & 1 & 0 & 0 & 0 \\ 0 & -1 & 1 & 1 & 0 \\ 0 & 0 & -1 & -1 & 1 \end{bmatrix} \end{matrix}$$

As it can be seen from Fig. 1. and the described matrices, the adjacency matrix "**A**" cannot completely describe the system, because it is not clear that nodes "2" and "3" are connected by "I2" or "I3". That is why we must use incidence matrix "**B**" for the complete definition of this graph.

From all this, we can conclude, if in the graph two paired nodes are directly connected with more than one edge (*parallel edge connections*), for mathematical representation the adjacency matrix is insufficient. It is recommended to use an incidence matrix.

2. Graph Creation and Segmentation

In this section the author will refer to his previously published papers and articles, that will develop more to the graphs, see appendix A, and [2]. The system in appendix A can be segmented from several aspects. Based on [3], some hierarchy can be stated, where the higher level is created from *modules*, the middle level is consisting of *units,* and the lower level includes the *sensors* and *actuators*. The hierarchy is important regarding connections. Between the high and middle level of hierarchy are *bi-directional* connections, while between the middle and low-level hierarchy are "one-way" connections. The orientation depends on endpoints. In case of actuators the orientation shows outwards from the units, whereas in case of sensors the direction is opposite.

Another aspect of segmentation of the network can be stated based on functionality of the devices. Thus it, is possible to make groups of devices which are systematically connected [2]. Based on this aspect we can distinguish six sub-systems, see appendix A (driver control system, body exterior system, …). The systematization creates the logical blocks of the network, -*logical* (functional) *segmentation.*

The physical segmentation can be down based on modularity, where the segments are created based on real modules, e.g.: segment-PDM will be created from devices which are physically connected to the PDM module, segment-BCM is created similarly, and thus creating a *physical segmentation* of the system.

The problem of these modularizations are, in both cases, the different modules are overlapping and the created graph will be quite complex, but it shows "the beauty" of the task. The starting point for graph creation has been solved in [3], where the author had to use some simplifications for a more complex representation. In the whole network, 5 *modules* can be found, which will be numbered as $M_{\{1,2,3,4,5\}}$; 10 *units* $U_{\{1,2,3,4,5,6,7,8,9,10\}}$; 20 *sensors* $S_{\{1,...,20\}}$ 18 *actuators* signed by $A_{\{1,...,18\}}$, and a Battery signed by B. These can be imagined as nodes of the graph. The edges will be noted by $\mathbf{e}_{1,..n}$. The starting figure for this marking can be found in [3].

For better understanding and a clearer graph, the author creates the group of sensors, resp. actuators, belonging to each device (modules and units). The *VSP* unit will be neglected, because in newer releases (newest version of Leaf Z0) it was no longer installed by the manufacturer. By this way the number of units decrease by one ($U_{\{1,..,9\}}$), and two of the actuators are also eliminated ($A_{\{1,..,16\}}$), and certainly the number of nodes also, and the edges also will be reduced, see Fig.2.

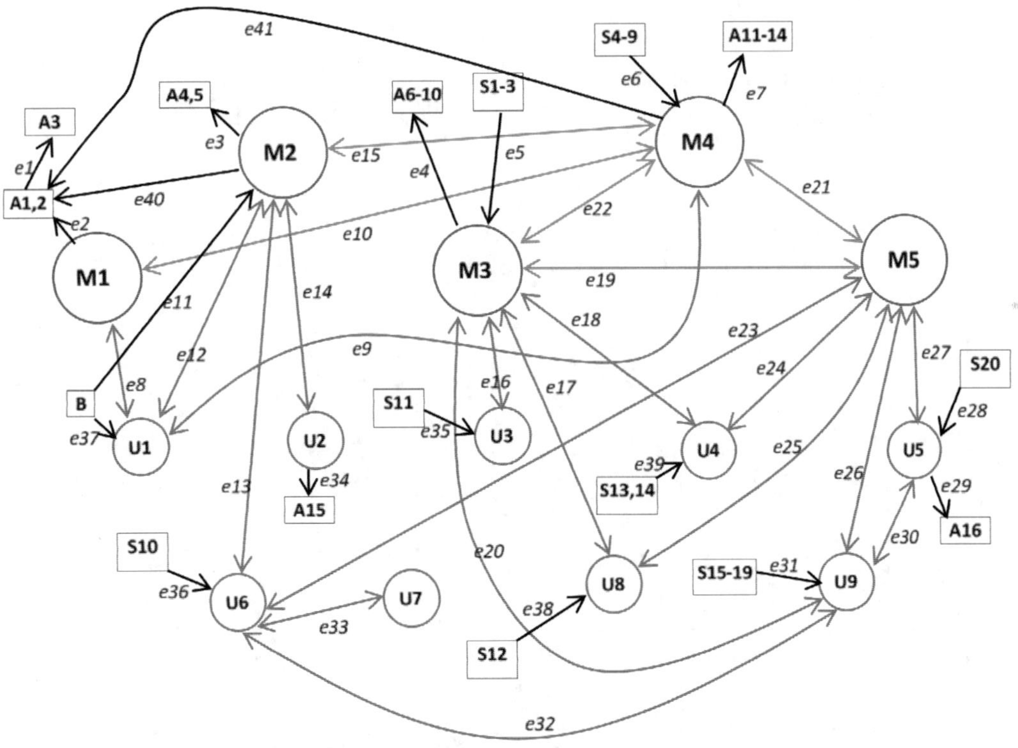

Figure 2. The simplified graph representation of the **Figure A1** in appendix A.

As it can be seen from Fig. 2. the oriented graph has **30** nodes composed from modules ($M_{(i)}$), units ($U_{(j)}$), sensors ($S_{(k)}$) and actuators ($A_{(l)}$), and **41** edges ($e_{(m)}$). These will create the rows and columns of the matrix.

Only for the sake of accuracy the non-simplified graph contains 54 nodes and 65 edges. Because of the limited boundries of the paper, the author will mathematically only represent the simplified graph.

3. Mathematical representation and matrix creation

As it can be seen that the graph is oriented. The orientation in the matrix will be represented as follows:

Let me define the incidence matrix by $\boldsymbol{B}_{(30x41)}=[b_{i,j}]$, where rows are the nodes (vertices) $v_{(i)}$ and the columns are the edges $e_{(j)}$. Then:

$$b_{i,j}\begin{cases}1 & v_i \underset{e_j}{\rightarrow} \\ -1 & v_i \underset{e_j}{\leftarrow} \\ \pm 1 & v_i \underset{e_j}{\leftrightarrow} \\ 0 & otherwise\end{cases}; \tag{3}$$

The grouping of the sensors and actuators is detailed in Table 1. and Table 2.

Table 1. Group of actuators and associated devices

groups:	A1	A2	A3	A4	A5	A6	A7
belonging actuators	*A1,2*	*A3*	*A4,5*	*A6-10*	*A11-14*	*A15*	*A16*

Table 2. Group of sensors and associated devices

groups:	S1	S2	S3	S4	S5	S6	S7	S8
belonging sensors	*S1-3*	*S4-9*	*S10*	*S11*	*S12*	*S13,14*	*S15-19*	*S20*

The simplified *incidence* matrix of the system, where the first row contains the number of edges, and the first column contains the main control units (M_{1-5}), units (U_{1-9}) and group of the sensors (S_{1-8}) and actuators (A_{1-7}).

	1	2	3	4	5	6	7	8	9	10	11	12	13	14	15	16	17	18	19	20	21	22	23	24	25	26	27	28	29	30	31	32	33	34	35	36	37	38	39	40	41
M1	0	1	0	0	0	0	0	±1	0	±1	0	0	0	0	0	0	0	0	0	0	0	0	0	0	0	0	0	0	0	0	0	0	0	0	0	0	0	0	0	0	0
M2	0	0	1	0	0	0	0	0	0	0	−1	±1	±1	±1	±1	0	0	0	0	0	0	0	0	0	0	0	0	0	0	0	0	0	0	0	0	0	0	0	0	1	0
M3	0	0	0	1	−1	0	0	0	0	0	0	0	0	0	0	±1	±1	±1	±1	±1	0	±1	0	0	0	0	0	0	0	0	0	0	0	0	0	0	0	0	0	0	0
M4	0	0	0	0	±1	−1	1	0	±1	0	0	0	0	0	±1	0	0	0	0	0	±1	±1	0	0	0	0	0	0	0	0	0	0	0	0	0	0	0	0	0	0	1
M5	0	0	0	0	0	0	0	0	0	0	0	0	0	0	0	0	0	0	±1	0	±1	0	±1	±1	±1	±1	±1	0	0	0	0	0	0	0	0	0	0	0	0	0	0
U1	0	0	0	0	0	0	0	±1	±1	0	0	±1	0	0	0	0	0	0	0	0	0	0	0	0	0	0	0	0	0	0	0	0	0	0	0	0	−1	0	0	0	0
U2	0	0	0	0	0	0	0	0	0	0	0	0	0	±1	0	0	0	0	0	0	0	0	0	0	0	0	0	0	0	0	0	0	0	1	0	0	0	0	0	0	0
U3	0	0	0	0	0	0	0	0	0	0	0	0	0	0	0	±1	0	0	0	0	0	0	0	0	0	0	0	0	0	0	0	0	0	0	1	0	0	0	0	0	0
U4	0	0	0	0	0	0	0	0	0	0	0	0	0	0	0	0	0	±1	0	0	0	0	0	±1	0	0	0	0	0	0	0	0	0	0	0	0	0	0	−1	0	0
U5	0	0	0	0	0	0	0	0	0	0	0	0	0	0	0	0	0	0	0	0	0	0	0	0	0	0	±1	−1	1	±1	0	0	0	0	0	0	0	0	0	0	0
U6	0	0	0	0	0	0	0	0	0	0	0	0	±1	0	0	0	0	0	0	0	0	0	±1	0	0	0	0	0	0	0	0	±1	±1	0	0	−1	0	0	0	0	0
U7	0	0	0	0	0	0	0	0	0	0	0	0	0	0	0	0	0	0	0	0	0	0	0	0	0	0	0	0	0	0	0	0	±1	0	0	0	0	0	0	0	0
U8	0	0	0	0	0	0	0	0	0	0	0	0	0	0	0	0	±1	0	0	0	0	0	0	0	±1	0	0	0	0	0	0	0	0	0	0	0	0	−1	0	0	0
U9	0	0	0	0	0	0	0	0	0	0	0	0	0	0	0	0	0	0	0	±1	0	0	0	0	0	±1	0	0	0	±1	−1	±1	0	0	0	0	0	0	0	0	0
A1	1	−1	0	0	0	0	0	0	0	0	0	0	0	0	0	0	0	0	0	0	0	0	0	0	0	0	0	0	0	0	0	0	0	0	0	0	0	0	0	−1	−1
A2	−1	0	0	0	0	0	0	0	0	0	0	0	0	0	0	0	0	0	0	0	0	0	0	0	0	0	0	0	0	0	0	0	0	0	0	0	0	0	0	0	0
A3	0	0	−1	0	0	0	0	0	0	0	0	0	0	0	0	0	0	0	0	0	0	0	0	0	0	0	0	0	0	0	0	0	0	0	0	0	0	0	0	0	0
A4	0	0	0	−1	0	0	0	0	0	0	0	0	0	0	0	0	0	0	0	0	0	0	0	0	0	0	0	0	0	0	0	0	0	0	0	0	0	0	0	0	0
A5	0	0	0	0	0	0	−1	0	0	0	0	0	0	0	0	0	0	0	0	0	0	0	0	0	0	0	0	0	0	0	0	0	0	0	0	0	0	0	0	0	0
A6	0	0	0	0	0	0	0	0	0	0	0	0	0	0	−1	0	0	0	0	0	0	0	0	0	0	0	0	0	0	0	0	0	0	0	0	0	0	0	0	0	0
A7	0	0	0	0	0	0	0	0	0	0	0	0	0	0	0	0	0	0	0	0	0	0	0	0	0	0	0	0	−1	0	0	0	0	0	0	0	0	0	0	0	0
S1	0	0	0	0	1	0	0	0	0	0	0	0	0	0	0	0	0	0	0	0	0	0	0	0	0	0	0	0	0	0	0	0	0	0	0	0	0	0	0	0	0
S2	0	0	0	0	0	1	0	0	0	0	0	0	0	0	0	0	0	0	0	0	0	0	0	0	0	0	0	0	0	0	0	0	0	0	0	0	0	0	0	0	0
S3	0	0	0	0	0	0	0	0	0	0	0	0	0	0	0	0	0	0	0	0	0	0	0	0	0	0	0	0	0	0	0	0	0	0	0	1	0	0	0	0	0
S4	0	0	0	0	0	0	0	0	0	0	0	0	0	0	0	0	0	0	0	0	0	0	0	0	0	0	0	0	0	0	0	0	0	0	1	0	0	0	0	0	0
S5	0	0	0	0	0	0	0	0	0	0	0	0	0	0	0	0	0	0	0	0	0	0	0	0	0	0	0	0	0	0	0	0	0	0	0	0	0	1	0	0	0
S6	0	0	0	0	0	0	0	0	0	0	0	0	0	0	0	0	0	0	0	0	0	0	0	0	0	0	0	0	0	0	0	0	0	0	0	0	0	0	1	0	0
S7	0	0	0	0	0	0	0	0	0	0	0	0	0	0	0	0	0	0	0	0	0	0	0	0	0	0	0	0	0	0	1	0	0	0	0	0	0	0	0	0	0
S8	0	0	0	0	0	0	0	0	0	0	0	0	0	0	0	0	0	0	0	0	0	0	0	0	0	0	0	1	0	0	0	0	0	0	0	0	0	0	0	0	0
B	0	0	0	0	0	0	0	0	0	0	1	0	0	0	0	0	0	0	0	0	0	0	0	0	0	0	0	0	0	0	0	0	0	0	0	0	1	0	0	0	0

Since the graph does not contain multi-edges assigned to the same pair of nodes, it is possible to write up the *adjacency* matrix $\boldsymbol{A}_{(30x30)}=[a_{i,j}]$ of the system too. For an adjacency matrix creation, I will use the definition stated in equation (3), because it makes no sense to use the definition stated in (1), since the sensors are connected to the modules representing a real connection between the devices. Only the direction of the data flow is opposite. It makes no sense to represent them by "0".

	M1	M2	M3	M4	M5	U1	U2	U3	U4	U5	U6	U7	U8	U9	A1	A2	A3	A4	A5	A6	A7	S1	S2	S3	S4	S5	S6	S7	S8	B
M1	0	0	0	±1	0	±1	0	0	0	0	0	0	0	0	1	0	0	0	0	0	0	0	0	0	0	0	0	0	0	0
M2	0	0	0	±1	0	±1	±1	0	0	0	±1	0	0	0	1	0	1	0	0	0	0	0	0	0	0	0	0	0	0	−1
M3	0	0	0	±1	±1	0	0	±1	±1	0	0	0	±1	±1	0	0	0	1	0	0	0	−1	0	0	0	0	0	0	0	0
M4	±1	±1	±1	0	±1	±1	0	0	0	0	0	0	0	0	1	0	0	0	1	0	0	0	−1	0	0	0	0	0	0	0
M5	0	0	±1	±1	0	0	0	0	±1	±1	±1	0	±1	±1	0	0	0	0	0	0	0	0	0	0	0	0	0	0	0	0
U1	±1	±1	0	±1	0	0	0	0	0	0	0	0	0	0	0	0	0	0	0	0	0	0	0	0	0	0	0	0	0	−1
U2	0	±1	0	0	0	0	0	0	0	0	0	0	0	0	0	0	0	0	0	1	0	0	0	0	0	0	0	0	0	0
U3	0	0	±1	0	0	0	0	0	0	0	0	0	0	0	0	0	0	0	0	0	0	0	0	0	−1	0	0	0	0	0
U4	0	0	±1	0	±1	0	0	0	0	0	0	0	0	0	0	0	0	0	0	0	0	0	0	0	0	0	−1	0	0	0
U5	0	0	0	0	±1	0	0	0	0	0	0	0	0	±1	0	0	0	0	0	0	1	0	0	0	0	0	0	0	−1	0
U6	0	±1	0	0	±1	0	0	0	0	0	0	±1	0	±1	0	0	0	0	0	0	0	0	0	−1	0	0	0	0	0	0
U7	0	0	0	0	0	0	0	0	0	0	±1	0	0	0	0	0	0	0	0	0	0	0	0	0	0	0	0	0	0	0
U8	0	0	±1	0	±1	0	0	0	0	0	0	0	0	0	0	0	0	0	0	0	0	0	0	0	0	−1	0	0	0	0
U9	0	0	±1	0	±1	0	0	0	0	±1	±1	0	0	0	0	0	0	0	0	0	0	0	0	0	0	0	0	−1	0	0
A1	−1	−1	0	−1	0	0	0	0	0	0	0	0	0	0	0	1	0	0	0	0	0	0	0	0	0	0	0	0	0	0
A2	0	0	0	0	0	0	0	0	0	0	0	0	0	0	−1	0	0	0	0	0	0	0	0	0	0	0	0	0	0	0
A3	0	−1	0	0	0	0	0	0	0	0	0	0	0	0	0	0	0	0	0	0	0	0	0	0	0	0	0	0	0	0
A4	0	0	−1	0	0	0	0	0	0	0	0	0	0	0	0	0	0	0	0	0	0	0	0	0	0	0	0	0	0	0
A5	0	0	0	−1	0	0	0	0	0	0	0	0	0	0	0	0	0	0	0	0	0	0	0	0	0	0	0	0	0	0
A6	0	0	0	0	0	0	−1	0	0	0	0	0	0	0	0	0	0	0	0	0	0	0	0	0	0	0	0	0	0	0
A7	0	0	0	0	0	0	0	0	0	−1	0	0	0	0	0	0	0	0	0	0	0	0	0	0	0	0	0	0	0	0
S1	0	0	1	0	0	0	0	0	0	0	0	0	0	0	0	0	0	0	0	0	0	0	0	0	0	0	0	0	0	0
S2	0	0	0	1	0	0	0	0	0	0	0	0	0	0	0	0	0	0	0	0	0	0	0	0	0	0	0	0	0	0
S3	0	0	0	0	0	0	0	0	0	0	1	0	0	0	0	0	0	0	0	0	0	0	0	0	0	0	0	0	0	0
S4	0	0	0	0	0	0	0	1	0	0	0	0	0	0	0	0	0	0	0	0	0	0	0	0	0	0	0	0	0	0
S5	0	0	0	0	0	0	0	0	0	0	0	0	1	0	0	0	0	0	0	0	0	0	0	0	0	0	0	0	0	0
S6	0	0	0	0	0	0	0	0	1	0	0	0	0	0	0	0	0	0	0	0	0	0	0	0	0	0	0	0	0	0
S7	0	0	0	0	0	0	0	0	0	0	0	0	0	1	0	0	0	0	0	0	0	0	0	0	0	0	0	0	0	0
S8	0	0	0	0	0	0	0	0	0	1	0	0	0	0	0	0	0	0	0	0	0	0	0	0	0	0	0	0	0	0
B	0	1	0	0	0	1	0	0	0	0	0	0	0	0	0	0	0	0	0	0	0	0	0	0	0	0	0	0	0	0

If the segmentation rules, mentioned above in *Section 2.*, for all of sub-systems will be enforced, then it is possible to create the sub-graphs/submatrices of the whole system. Just as an example let's have a look to the sub-system connected to module "M1". Module M1 is connected to the vertices M4, U1, A1, A2. The created sub-graph can be seen on Fig. 3.

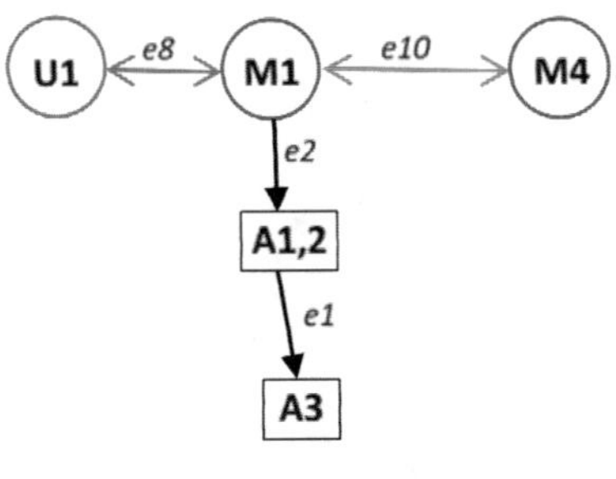

Figure 3. The sub-graph of "M1" module

And now the appropriate adjacency ($\boldsymbol{M1A}_{(5x5)}$) and incidence ($\boldsymbol{M1I}_{(5x4)}$) matrices of the subsystem.

$$M1A_{(5x5)} = \begin{array}{c|ccccc} & M1 & M4 & U1 & A1 & A2 \\ M1 & 0 & \pm1 & \pm1 & 1 & 0 \\ M4 & \pm1 & 0 & 0 & 0 & 0 \\ U1 & \pm1 & 0 & 0 & 0 & 0 \\ A1 & -1 & 0 & 0 & 0 & 1 \\ A2 & 0 & 0 & 0 & -1 & 0 \end{array} \qquad M1I_{(5x4)} = \begin{array}{c|cccc} & e1 & e2 & e8 & e10 \\ M1 & 0 & 0 & \pm1 & \pm1 \\ M4 & 0 & 0 & 0 & \pm1 \\ U1 & 0 & 0 & \pm1 & 0 \\ A1 & 0 & -1 & 0 & 0 \\ A2 & -1 & 0 & 0 & 0 \end{array} \tag{4}$$

On the other hand, in the case that we want to examine the whole system more precisely, on how the individual elements interact, it is unavoidable to describe the whole matrices.

Conclusion

This paper should solve for further research as can be seen in [5], [6], [7]. Furthermore, it showed an example of how the system can be segmented and created an incidence matrix of a simplified electronic system of a given vehicle. Why should this be useful? Usually in real life, instead of the

value of matrix elements (1, 0, ±1), it is possible to write the required voltage level or current on the selected edge (circuit branch). The created matrix can be realized by some FPGA circuit, which can control (check) all of the connections during operation (real-time), and in case of improper data on the selected link some action can be initiated. By this way a "log-file" can be created, that logs the errors, events (by timestamp or stamp, based on traveled kilometers), that have occurred during operation.
Further possible use of the system can be connected to a risk analysis, in the event of a system failure. Here can be examined, how the failure is spreading in the system and how this error will influence the sub-parts of the system.
The nomenclature of abbreviations used in **Figure A1.**, can be found in reference [3]. The abbreviations used in this article can be found in **appendix B.**

Future Work

In case of mathematical representations of the sub-systems, the layered hyper-matrix should be used, where the connections between the different submatrices can be displayed. The theoretical architecture of this hyper-matrix can be seen on Figure 4., where $\boldsymbol{CM}_{(i,j)}$ matrices are the layered sub-matrices of the system.

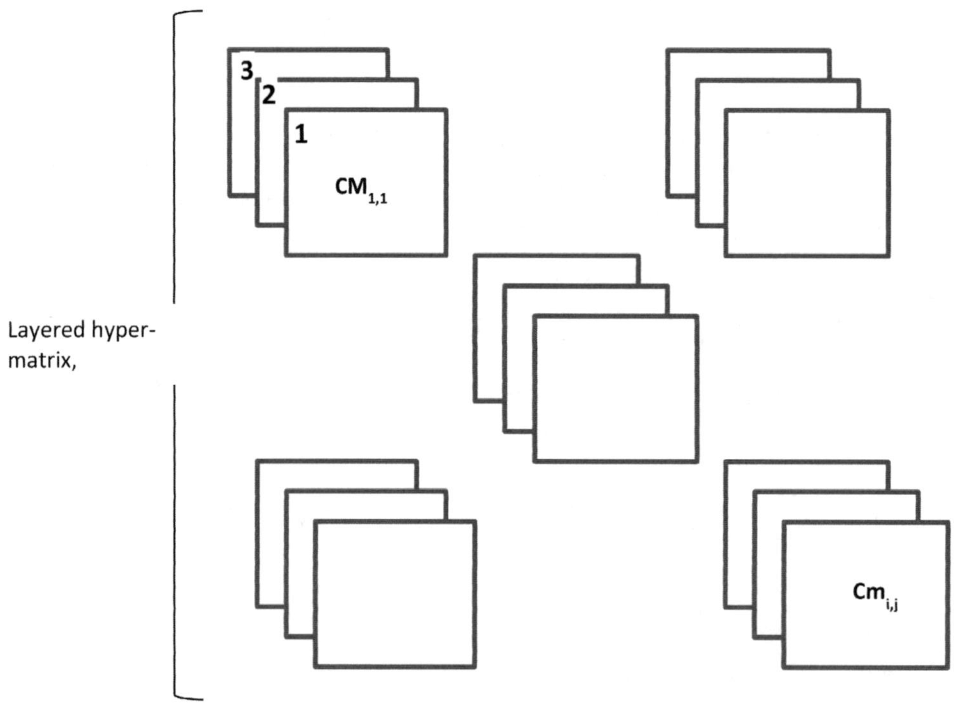

$$H = \begin{bmatrix} \begin{bmatrix} CM(1,1)_{1,1} & \cdots & CM(1,1)_{1,n} \\ \vdots & \ddots & \vdots \\ CM(1,1)_{m,1} & \cdots & CM(1,1)_{m,n} \end{bmatrix} & \cdots & [CM(1,J)_{m,n}] \\ \vdots & \ddots & \vdots \\ [CM(I,1)_{m,n}] & \cdots & \begin{bmatrix} CM(I,J)_{1,1} & \cdots & CM(I,J)_{1,n} \\ \vdots & \ddots & \vdots \\ CM(I,J)_{m,1} & \cdots & CM(I,J)_{m,n} \end{bmatrix} \end{bmatrix}$$

Figure 4. Possible solving of the overlapped sub-parts of the system

The question is, how many layers should be formed? Only at the idea level, I could suggest, that so many sub-matrices, as many sub-systems. One sub-matrix contains the elements of the sub-system, and if any element is connected to more sub-systems; here is reason of using the layer of the matrix. In this case, from Figure 2., it can be seen, that actuator group A1,2 is connected simultaneously to module M1, and M2 too. Here, the first layer of M1 can be created from modules, which are connected only to the M1, and the additional layers would include those elements that are also connected to the other modules.

References

[1] Nagy I and Vér F.B. 2019 Practical Analysis of an AMR Wheel Speed Sensor and CAN Bus System for the Communication's Failures and its Implications to the EV's Control System, *Proc. of the Sisy 2019*, pp 241-247.

[2] Nagy I 2019 Hierarchical Mapping of an Electric Vehicle Sensor and Control Network; *Acta Polytechnica Hungarica,* **18** : 9 pp.: 161-180.

[3] Nagy I 2021 Sensors Classification and Hierarchical Organization of Control and Communication System in an Electric Vehicle; *Proc. of SACI*2021, pp. 287-292, doi: 10.1109/SACI51354.2021.9465638.

[4] Heinold B 2020 *A Simple Introduction to Graph Theory*; e-Book; Licensed under a Creative Commons Attribution-Noncommercial-Share Alike 3.0 Unported License, last update: september 23, 2020.

[5] Pokorádi L 2018 Methodology of Advanced Graph Model-based Vehicle Systems' Analysis; in *Proc of CINTI2018* Budapest Hungary pp.: 325-328.

[6] Pokorádi L 2008 *Rendszerek és folyamatok modellezése* (Campus kiadó, Debrecen, Hungary)

[7] Pokorádi L 2018 Graph model-based analysis of technical systems, *MATERIALS SCIENCE AND ENGINEERING* 393 : 1 Paper: 012007 , 8 p.

[8] Attenborough M 2003 *Mathematics for Electrical Engineering and Computing* **19** Graph Theory, pp.: 461-478, (Newness, Elsevier) ISBN 9780750658553.

IOP Conf. Series: Materials Science and Engineering 1237 (2022) 012003 doi:10.1088/1757-899X/1237/1/012003

appendix A.

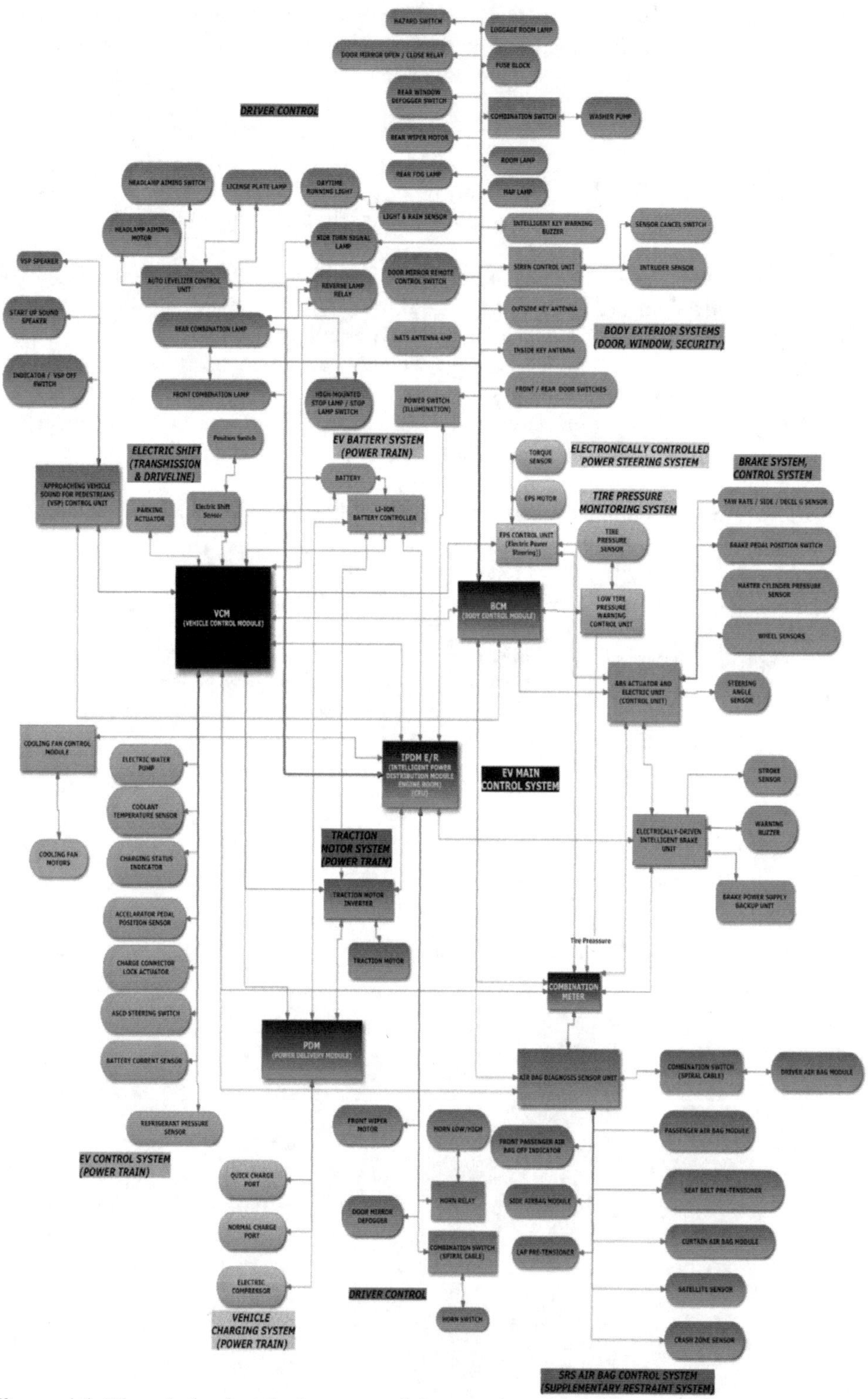

Figure A1. The whole electrical system of the mentioned vehicle, see footnote 1, and [2].

IOP Conf. Series: Materials Science and Engineering 1237 (2022) 012003 doi:10.1088/1757-899X/1237/1/012003

appendix B.

Table of groups of modules, units, sensors and actuators, where the different groups are labelled by different colours in the "Nr." column. The obvious acronyms are not explained.

Table B1. Groups of the simplified units and assigned modules/units/sensors/actuators

Nr.	Symbol	Acronym	Meaning
1	M1	PDM	Power Delivery Module
2	M2	IPDM E/R	Intelligent Power Distribution Module, Engine Room
3	M3	BCM	Body Control Module
4	M4	VCM (VC)	Vehicle Control Module
5	M5	COM. METER	Combination Meter
6	U1	Li-Ion Bat. Unit	
7	U2	Cooling Fan Unit	
8	U3	Siren Control	
9	U4	Air Bag Diagnostic Control	
10	U5	EPS Control	Electric Power Steering System
11	U6	E-d Intelligent	Electrically Driven Intelligent Brake Unit
12	U7	Brake P.S. Backup	Brake Power Supply Backup unit
13	U8	Low TPMS Warning Unit	TPMS-Tyre Pressure Monitoring System
14	U9	ABS Control	Anti-lock Braking System
15	A1	Electric Compressor	
16	A2	Traction Motor Inverter	
17	A3	Traction Motor	
18	A4	Wiper Motors	
19	A5	Door/Mirror Defogger	
20	A6	Power Sw Illuminator	Power Switch Illuminator
21	A7	Lamps: Daytime/ Fog/Room/Luggage	
22	A8	Door Sw.: front/Rear/Tail/Hoods	Door Switch
23	A9	iKey warning buzzer	intelligent Key Warning Buzzer
24	A10	Washer pumps: front/rear	
25	A11	Charge Connector, Lock actuator	
26	A12	Parking Actuator	
27	A13	Electric Water Pump	
28	A14	Reverse Lamp Relay	
29	A15	Cooling Fun Motors	
30	A16	EPS Motor	Electric Power Steering System Motor
31	S1	NATS Antennas	Nissan Anti- Theft System
32	S2	Door/Mirror RC Sw	
33	S3	Light Rain Sensor	
34	S4	Electric Shift pos. Sensor	
35	S5	ASCD Steering Sw	Automatic Speed Control Device steering Switch
36	S6	Coolant Temperature Sensor	
37	S7	Accelerator Pedal Pos. Sensor	
38	S8	Battery Current Sensor	
39	S9	Refrigerant Pressure Sensor	
40	S10	Stroke Sensor	
41	S11	Intruder Sensor	
42	S12	Tyre pressure sensor	
43	S13	Satellite Sensor	
44	S14	Seat Belt Pretension Sensing	
45	S15	Steering Angle Sensor	
46	S16	Wheel Speed Sensor	
47	S17	Master Cylinder Pressure Sensor	
48	S18	Yaw/Side/Decel-G. Sensor	Yaw/Side Deceleration Gyro Sensor
49	S19	Brake Pedal pos. Sensor	
50	S20	Torque Sensor	
51	B	Battery	
		VSP	Vehicle Sound for Pedestrians (neglected)

IOP Conf. Series: Materials Science and Engineering 1237 (2022) 012004 doi:10.1088/1757-899X/1237/1/012004

Monte-Carlo Simulation-based Analysis of Bridge Structure Systems' Reliability

László Pokorádi

Óbuda University, Institute of Mechatronics and Vehicle Engineering,
Népszínház u. 8, H-1081 Budapest, Hungary

pokoradi.laszlo@bgk.uni-obuda.hu

Abstract. Modern engineering systems, such as automobiles, should have reliability requirements. This paper focuses on the reliability of System with Complex Interconnection (SwCI). Most of the real complex systems in the modern vehicles are different sensors, as Bridge Structure Systems (BSSs). The BSSs are used as an example in the field of system reliability analysis. The proposed Monte-Carlo Simulation-based investigation can be used for the estimation of the Required Number of Spare Parts (RNSPs).

1. Introduction

The reliability analysis of technical systems and networks is a very important tool in analyzing and quantifying technical safety.

The objective of Labašova's paper was an overview to determine the reliability of Systems with Simple Interconnections (SwSI) [1]. Her aim was to analyze the reliability of the systems with parallel or serial arrangements too. According to Labašova, the Reliability Block Diagram (RBD) method of reliability is relatively simple to process. For less complex systems, it does not require software support, and is therefore relatively widely used. The systems that do not have so-called simple connections are named Systems with Complex Interconnection (SwCI) in other word: complex systems [2]. During the investigation of reliabilities of complex systems, they cannot be simplified to single equivalent blocks by a combination of series and/or parallel reductions. Therefore, the methods of RBD cannot be able to determine the reliability of the SwCI.

Most of the concrete complex systems in the modern vehicle industry are vehicle sensors, as a Bridge Structure System (BSS), and as sensory networks [3]. Nagy has investigated the electronic control network of a particular vehicle [4]. The mapped system can provide a base for evaluation of system reliability or for analysis of error-spreading on the network.

In addition to vehicle sensors, the capacity and safety of transport systems can also be analyzed as SwCI. Péter's research work is aimed to develop applicable mathematical methods that accurately consider the characteristics of the trajectories. The aim of his paper [5] was to develop an efficient and fast algorithm that resulted in the determination of all the different trajectories.

In Cristescu's publication, new techniques of approximation of the reliability of a two-terminal network are worked out based on the constructive theory of functions and few related methods [6].

Mathematical modeling is the representations of the process occurring in given modeled technical equipment by symbols or symbol systems of mathematics, from the point of view of the given investigation. Nowadays real engineering systems consist of a great number of interconnected aggregates. The mathematical model is a simplified representation of a real physical system or technical process; thus, it can be imprecise and accordingly it has uncertainties.

If the modeler fails to reduce the model properly, the uncertainty is called epistemic.

The other type of uncertainty is approximated with the model-based investigated system or its neighbourhood and it is named parametric.

This parametric uncertainty can be investigated using the Monte-Carlo Simulation (MCS) [7]. MCS is one of the classical simulation techniques. In their paper, Metropolis and Ulam named the method Monte-Carlo [8]. This publication served as motivation and offered the general description of a method dealing with a class of problems in mathematical physics.

The main object of this paper is to work out a probabilistic uncertainty analysis method of BSSs' reliability using MCS.

The paper is organized as follows: Section 2 shows the methodology of calculation of the reliability of BSS. Section 3 presents the MCS-based analysis of probability uncertainties of reliability of BSS, in the cases of different situations. Section 4 summarizes the paper and its conclusions, then outlines the future scientific work of the Author.

2. Reliability of the Bridge Structure System

The components of canonic systems have only two states – work (it performs its required function – designated as "+") and wrong (it does not perform the required function – designated by "-") [9].

A BSS is composed of five elements. We assumed that the individual blocks (and the system) are independent (the state of any block does not affect the state of other ones in any way) and two-state. Their availabilities can be characterized by probability of its failure p_i and its reliability r_i . Their sum must be 1:

$$r_i + p_i = 1 \quad i \in [A;B;C;D;E] \tag{1}$$

One of the applicable methods is to determine the reliability of BSS by the sum of the probabilities of an investigated system, system states.

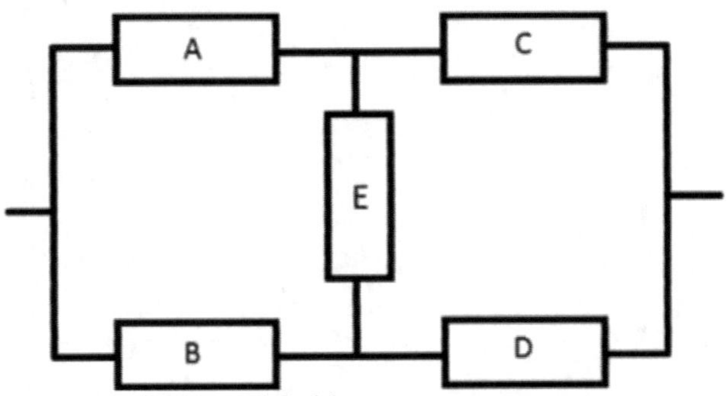

Figure 1 Bridge structure system

The possible system states of the BSS are summarized in the form of a Truth Table, shown in Table 1, with each block assigned either a good or a faulty state. The Q_i column includes the probabilities of each of the possible system states.

2.1. General Case

In the general case a BSS works if a matter, sign or energy "can cross" the system.

The state probabilities resulting in an operating system are included in the rows 1; 2; 3; 5; 6; 7; 9; 10; 11; 17; 18; 19; 21; 22; 25; and 27. The sum of the operating system state probabilities

$$R_{sys} = Q_1 + Q_2 + Q_3 + Q_5 + Q_6 + Q_7 + Q_9 + Q_{10} + Q_{11} + Q_{17} + Q_{18} + Q_{19} + Q_{21} + Q_{22} + Q_{25} + Q_{27} \tag{2}$$

included in column Q_i is the reliability of the system.

The Table listing all possible states and their probabilities is called the Truth Table Method (TTM).

Table 1 Truth Table of BSS

j	A	B	C	D	E	System	Q_j
1	+	+	+	+	+	+	$r_A r_B r_C r_D r_E$
2	-	+	+	+	+	+	$(1-r_A) r_B r_C r_D r_E$
3	+	-	+	+	+	+	$r_A (1-r_B) r_C r_D r_E$
4	-	-	+	+	+	-	$(1-r_A)(1-r_B) r_C r_D r_E$
5	+	+	-	+	+	+	$r_A r_B (1-r_C) r_D r_E$
6	-	+	-	+	+	+	$(1-r_A) r_B (1-r_C) r_D r_E$
7	+	-	-	+	+	+	$r_A (1-r_B)(1-r_C) r_D r_E$
8	-	-	-	+	+	-	$(1-r_A)(1-r_B)(1-r_C) r_D r_E$
9	+	+	+	-	+	+	$r_A r_B r_C (1-r_D) r_E$
10	-	+	+	-	+	+	$(1-r_A) r_B r_C (1-r_D) r_E$
11	+	-	+	-	+	+	$r_A (1-r_B) r_C (1-r_D) r_E$
12	-	-	+	-	+	-	$(1-r_A)(1-r_B) r_C (1-r_D) r_E$
13	+	+	-	-	+	-	$r_A r_B (1-r_C)(1-r_D) r_E$
14	-	+	-	-	+	-	$(1-r_A) r_B (1-r_C)(1-r_D) r_E$
15	+	-	-	-	+	-	$r_A (1-r_B)(1-r_C)(1-r_D) r_E$
16	-	-	-	-	+	-	$(1-r_A)(1-r_B)(1-r_C)(1-r_D) r_E$
17	+	+	+	+	-	+	$r_A r_B r_C r_D (1-r_E)$
18	-	+	+	+	-	+	$(1-r_A) r_B r_C r_D (1-r_E)$
19	+	-	+	+	-	+	$r_A (1-r_B) r_C r_D (1-r_E)$
20	-	-	+	+	-	-	$(1-r_A)(1-r_B) r_C r_D (1-r_E)$
21	+	+	-	+	-	+	$r_A r_B (1-r_C) r_D (1-r_E)$
22	-	+	-	+	-	+	$(1-r_A) r_B (1-r_C) r_D (1-r_E)$
23	+	-	-	+	-	-	$r_A (1-r_B)(1-r_C) r_D (1-r_E)$
24	-	-	-	+	-	-	$(1-r_A)(1-r_B)(1-r_C) r_D (1-r_E)$
25	+	+	+	-	-	+	$r_A r_B r_C (1-r_D)(1-r_E)$
26	-	+	+	-	-	-	$(1-r_A) r_B r_C (1-r_D)(1-rE)$
27	+	-	+	-	-	+	$r_A (1-r_B) r_C (1-r_D)(1-r_E)$
28	-	-	+	-	-	-	$(1-r_A)(1-r_B) r_C (1-r_D)(1-r_E)$
29	+	+	-	-	-	-	$r_A r_B (1-r_C)(1-r_D)(1-r_E)$
30	-	+	-	-	-	-	$(1-r_A) r_B (1-r_C)(1-r_D)(1-r_E)$
31	+	-	-	-	-	-	$r_A (1-r_B)(1-r_C)(1-r_D)(1-r_E)$
32	-	-	-	-	-	-	$(1-r_A)(1-r_B)(1-r_C)(1-r_D)(1-r_E)$

2.2. *Wheatstone Like Bridge*

In the automotive industry, sensors usually are Wheatstone Bridges or Wheatstone Like Bridges (WLB). In the case of the working state of a sensor, its components should be in an operational state. So, the reliability of the WLB can be calculated as the probability of the system state 1 (see Table I):

$$R_{WLB} = Q_I = r_A r_B r_C r_D r_E \quad (3)$$

3. The Monte-Carlo Simulation

The steps of MCSs are the following:

1) Determination or estimation of density functions *f(x)* of inputs.
2) Specified-distribution excitation of model by inputs based on the results of step 1)
3) Statistical investigation of aoutput data collected from simulations conducted according to step 2).
4) Deduction of conclusions from the point of view of the engineering investigation.

3.1. Creation of Initial Data

As the first step, a statistical analysis was performed using available reliability data. The maximum, minimum, expected (mean) values and standard deviations, of block reliabilities were determined. The investigated time interval was 1 year (8760 hours). Due to the relatively small number of available data, the goodness-of-fit tests were left out. It was assumed that the measured parameters follow normal (Gauss) probability distribution.

Table 2 Statistical parameters of (measured) data

	Number of samples	Mean	St.Dev.	Minimum	Maximum
r_A	21	0.9401	0.00201	0.93032	0.94997
r_B	19	0.9703	0.00071	0.96671	0.97360
r_C	25	0.9398	0.00212	0.93047	0.94972
r_D	20	0.9503	0.00039	0.94879	0.95136
r_E	23	0.9104	0.01400	0.84079	0.97349

3.2. Simulation

During the simulation to generate actual values of reliability of blocks, the so-called Acceptance – Rejection method was used. Table 3 illustrates input data of the simulation based on the results of the statistical analysis.

Table 3 Input data of the simulation

	Mean	St.Dev
r_A	0.940	0.0020
r_B	0.970	0.0007
r_C	0.940	0.0021
r_D	0.950	0.0004
r_E	0.910	0.0140

Using the input data determined above, the reliability models as laid out in Section 2 were generated. During simulation the number of excitations was increased when the relative differences between actual and last result was more the *0.005*. Figures 2 – 5 depict mean values and standard deviations of system reliabilities depend on the number of excitations. The results can be seen In Table 4 when the number of excitations is one million.

Table 4 Results of Simulations

	Mean	St.Dev.
R_{sys}	0.993685	$1.5233\ 10^{-4}$
R_{WLB}	0.740941	0.012050

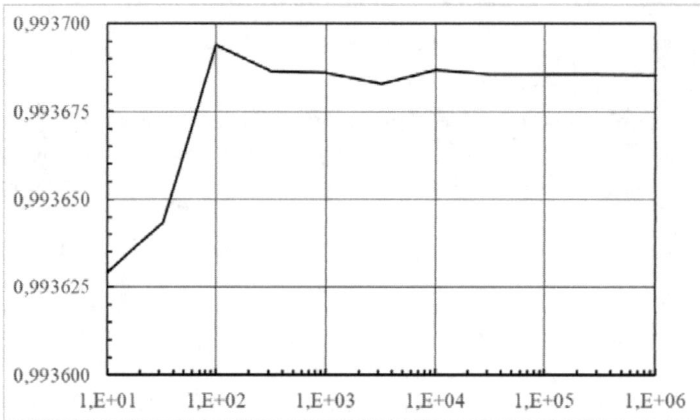

Figure 2 Mean Values of System Reliability in General Case

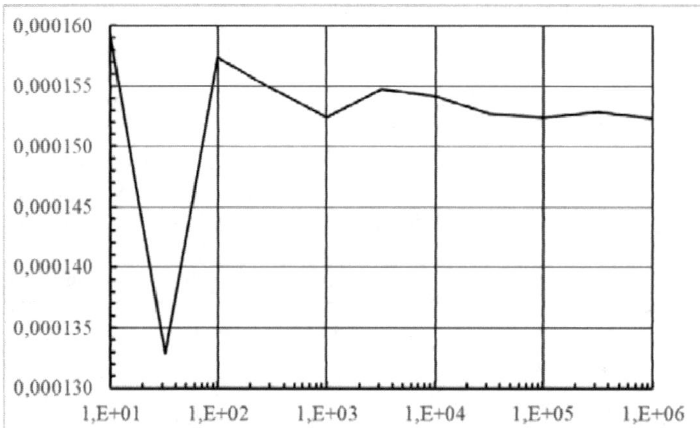

Figure 3 Standard Deviations of System Reliability in General Case

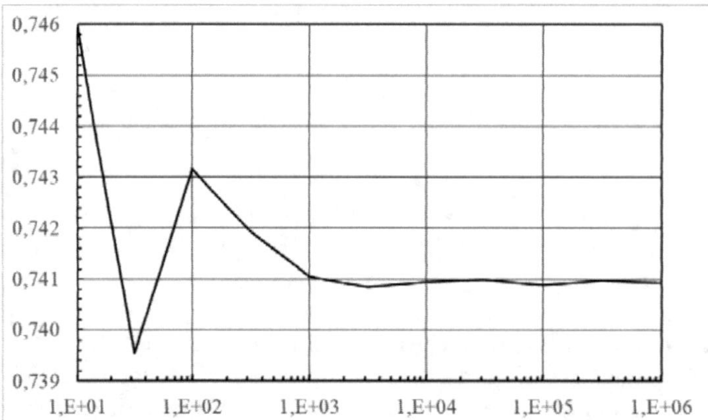

Figure 4 Mean Values of Wheatstone Like Bridge Reliability

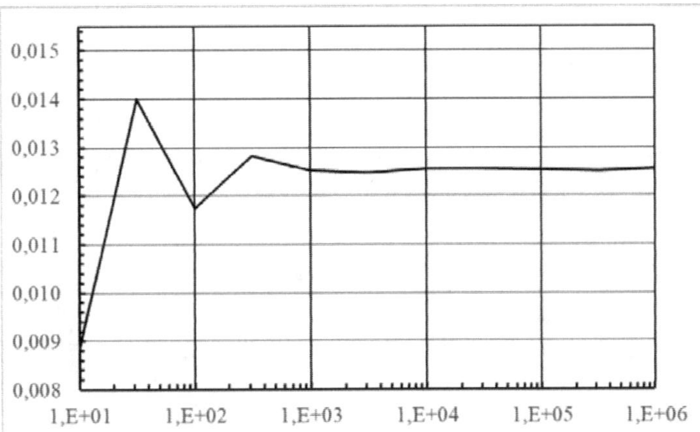

Figure 5 Standard Deviations of Wheatstone Like Bridge Reliability

4. Determination of Requested Number for Spare parts

One of the most important questions from maintenance management point of view is the Required Number of Spare Parts (RNSP). Knowing the reliability of the system R, the RNSP can be determined by the equation

$$RNSP = \left\lceil \left(\frac{1}{R} - 1 \right) N \right\rceil , \tag{4}$$

where N is the number of applied equipment (in this study: $N = 50000$).

For example, on the supposition that the uncertainty of system reliability follows a normal distribution; the probability of not having a spare part in the case of equipment failure is 0.05 and the following holds true:

$$R_{RNS} = m_R - 1.65 s_R \tag{5}$$

The RNSPs were determined for the cases of different assessing uncertainty values. These results are shown in Table 5.

Table 5 Required Number for Spare Part Depending on Estimating Uncertainty

Estimating Uncertainty	Number for Spare Part	
	General Case	WLB
10 %	328	18928
5 %	331	19343
2 %	334	19821
1 %	336	20152
0,5 %	338	20438
0,2 %	340	20798
0,01 %	342	21065

5. Correlation Analysis

The correlation coefficient characterizes the strength of linear stochastic interdependencies of the random variables.

If one of the two parameters having a strong negative correlation should change in some direction, another one will change in the opposite direction most probably according to the correlation coefficient.

The correlation coefficient $r_{\eta\mu}$ can be estimated empirically by the equation

$$r_{\eta\mu} = \frac{\sum_{i=1}^{n}\left(x_i - \sum_{j=1}^{n} x_j\right)\left(y_i - \sum_{j=1}^{n} y_j\right)}{\sqrt{\sum_{i=1}^{n}\left(x_i - \sum_{j=1}^{n} x_j\right)^2\left(y_i - \sum_{j=1}^{n} y_j\right)^2}} \quad (6)$$

using the samples x_1; x_2; … x_n and y_1; y_2; … ;y_n which belong to the variables η and μ [10].

The results of correlation analysis can be seen In Table 5 when the number of excitations is one million.

Table 6 Correlation Coefficients

	r_A	r_B	r_C	r_D	r_E
R_{sys}	0.4272	0.4668	0.6810	0.2299	0.2832
R_{WLB}	0,1312	0.0443	0.1372	0,260146	0,9461

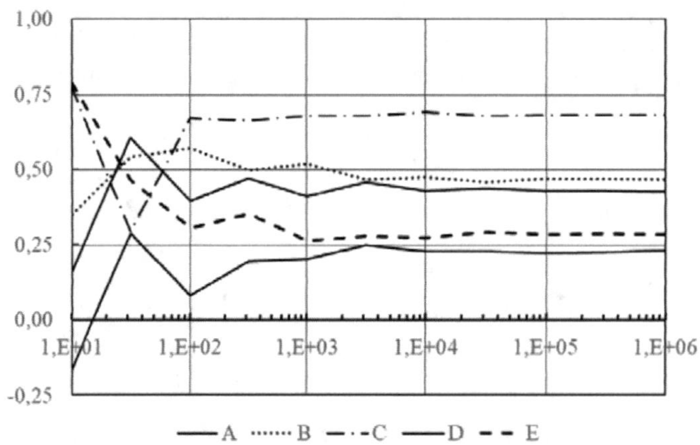

Figure 6 Correlation Coefficients in General Case

6. Conclusions

In this paper, a(n) MCS-based probabilistic uncertainty analysis method of BSSs' reliability was presented. Its possibilities of implementation have been shown by a practical case study of the BSSs' reliabilities.

The following conclusions can be deduced from the results of the analyses afore mentioned:

1) The proposed Monte-Carlo Simulation-based method can be used for investigating the uncertainty of system reliability and the Required Number for Spare Parts depending on the required estimating uncertainty.

2) The correlation analysis highlights

- the reliability of WLB has relative forceless correlations with the reliabilities of the elements A; B; C and D;

- the reliability of WLB has high positive correlations with reliability of block E, so improvement of E block reliability, results in increasing of reliability of WLB most significantly.

3) The disadvantage of the proposed TTM lies in the fact that the number of possible states of the system grows exponentially, depending on the number of blocks.

4) The drawback of MCS is that its elapsed time increases markedly, if the number of excitations increase (see Figure 8).

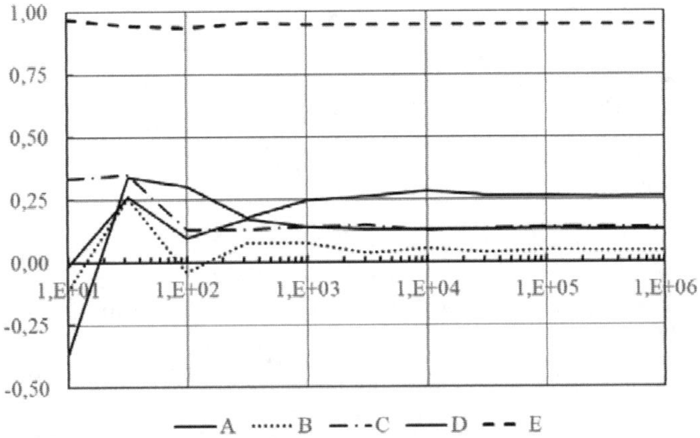

Figure 7 Correlation Coefficients of Wheatstone Like Bridge Reliability

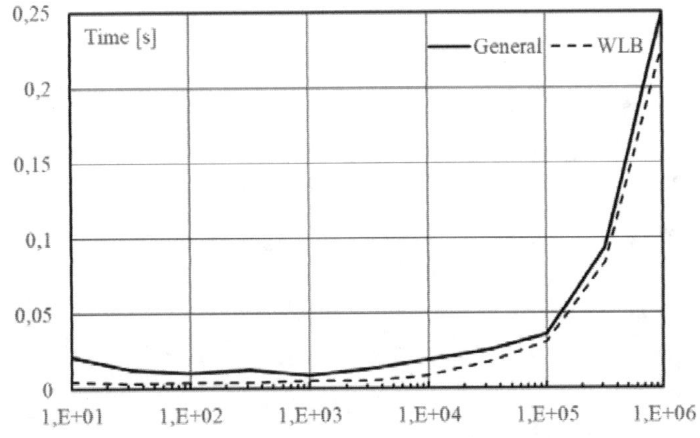

Figure 8 Elapsed Time of Monte-Carlo Simulation depends on Number of Excitations

5) The shortcoming of the proposed parametric uncertainty analysis lies in the fact that its result can be impossible from technical point of view.

In the case of the reliability modeling, one may find theoretically a reliability which is greater than 1 or less than 0; because the probability distributions generally have infinite domains from a theoretical (mathematical) point of view. This drawback can be eliminated by Probability–Bounds Analysis [11].

The Author's proposed prospective research direction in the future is the investigate of uncertainty analysis methodologies of SwCIs, such as vehicle sensory networks and sensors' reliability based on the Probability-Bounds Analysis.

References

[1] Labašová E. 2021 Reliability Assessment of the Systems with Simple Interconnections *Research Papers Faculty of Materials Science and Technology Slovak University of Technology* **29** 36-42
[2] Myers A 2010 Complex System Reliability (London: Springer)
[3] Pokorádi L. 2019 Interval Uncertainty Analysis of Reliability of Systems with Complex Interconnections GRADUS 6 145-155
[4] Nagy I. 2021 Hierarchical Mapping of an Electric Vehicle Sensor and Control Network *Acta Polytechnica Hungarica* **18** 161-180
[5] Péter T. et al. 2021 Analysis of Network Traversal and Qualification of the Testing Values of Trajectories *Acta Polytechnica Hungarica* **18** 151-171
[6] Cristescu, G., Drago V-F. 2021, Efficient Approximation of Two-Terminal Networks Reliability Polynomials Using Cubic Splines, *IEEE Transactions on Reliability*, **70** 1193-1203
[7] Pokorádi L 2014 Sensitivity analysis of reliability of Systems with Complex Interconnections *Journal of Loss Prevention in the Process Industries*, **32** 436-442
[8] Metropolis, N., Ulam, S. 1949 The Monte Carlo Method *Journal of the American Statistical Association* **44** 335-341
[9] Pokorádi L 2015 Failure probability analysis of bridge structure systems *Proc. of the SACI 2015* 147-150.
[10] Korn, T. A., Korn T. M. 2000 Mathematical Handbook for Scientists and Engineers (Dover Publication)
[11] Oberkampf WL, DeLand SM, Rutherford BM, Diegert, KV and Alvin KF. 2002 Error and uncertainty in modeling and simulation *Reliability Engineering and System Safety* **75** 333-357

The impact of COVID on the digitization of Hungarian maintenance sector

L Juhász[1] and L Pokorádi[2]

[1] PhD student, Doctoral School on Safety and Security Sciences, Óbuda University, Bécsi út 96/b, H-1034 Budapest, Hungary
[2] full professor, Óbuda University, Institute of Mechatronics and Vehicle Engineering, Népszínház u. 8, H-1081 Budapest, Hungary
j.laszlo92@stud.uni-obuda.hu

Abstract. One of the biggest recent innovation trends in the manufacturing industry is the use of Industry 4.0 tools and the introduction of various digitalization solutions. However, in 2020, the COVID19 pandemic significantly changed the way companies organize their day-to-day operations and the way they plan to innovate in their sector, in the near future. In this paper, we present the views of maintenance professionals in Hungarian automotive companies. We have compared the status of different types of enterprises. In particular, we looked at the differences in the status of these companies in terms of digitalization development. This allowed us to draw further conclusions.

1. Introduction

Maintenance is one of the most important processes that affect thruput for the production within a company [1]. Hungary's most technologically advanced manufacturing companies include car manufacturers, and their supply chain partners. In Hungary, the innovations that affect an entire company, usually appear first in the automotive industry. Moreover, the domestic sector has been characterized by a very high level of overall development in recent times. In 2019, the number of cars produced in Hungary reached a record high [2]. The sector has also seen very strong investment. Industry 4.0 and various industrial digitalization solutions were also seen as a further development point and innovation perspective by the Hungarian automotive industry players.

This great upward trajectory for the entire automotive industry was interrupted in the year 2020 by the COVID-19 pandemic [3] that also hit this country. The initial phase of the pandemic was characterized by supply chain disruptions [4], followed by a downward trend in new car sales [5]. A further major change in the life of companies was the increasing use of home offices in more and more workplaces in order to maintain social distancing [6]. In addition, the whole sector is facing a major technology shift, with manufacturers announcing successive target dates for the phasing out of internal combustion technology [7]. In our research, we seek to understand how the pandemic has affected the transition to new digital technologies by automotive players in the domestic industry, in particular maintenance departments.

The current paper presents the partial results of a larger, longer-term study. The primary results of this research have been previously reported in various studies [8][9][10]. The main steps of our larger research are illustrated in Figure 1. In this framework, we first defined the objectives of our research and then started a literature search in order to draw on the latest available expertise in the field for further research. As a result, we were able to publish several summary articles on the relationship between digital solutions and modern industrial maintenance [8][9]. This was followed up by qualitative research interviews with maintenance managers and professionals. For quantitative results, we conducted a preliminary survey of automotive industry players in Hungary [10]. Here, our aim was to get a general impression from the professionals about the current trends in the Hungarian sector and to explore the interrelationships between the trends. A further objective was to test our sampling and evaluation methodology. These results were used to prepare our final survey. The purpose of this paper is to present the results of this survey about the effects of the COVID-19 pandemic on the industry.

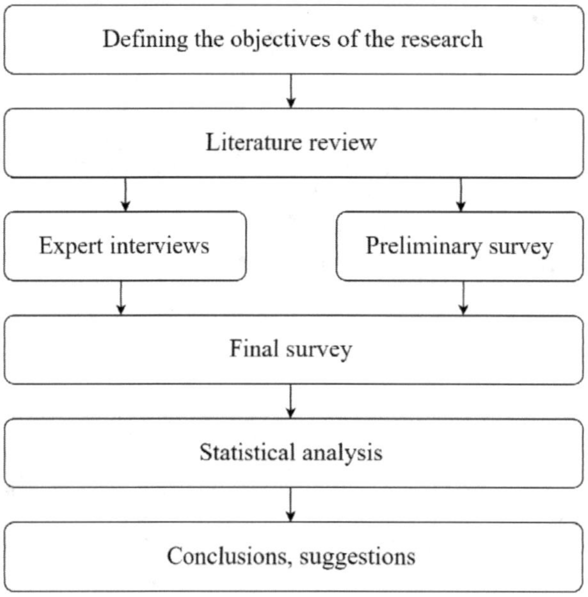

Figure 1. The main steps of the larger research (own figure).

The study consists of the following chapters: In Section 2, the survey methodology and the data collection process are described. In Section 3, the data provided by respondents will be analyzed. In Section 4, we describe the conclusions that can be drawn from the data. Finally, the authors summarize the study results and articulate their future research objectives.

2. Methodology of the survey

The research was conducted in the form of an online questionnaire. In this questionnaire, we also asked a number of additional questions, but only the points relevant to this publication are presented here. The survey started with questions on different individual parameters of the company in order to be able to draw appropriate conclusions for a subgroup. Based on our experience and previous research [11], there is a big difference in the work culture of domestic compared to foreign-owned companies and their views on different issues. For this reason, the nationality of ownership of the company was a particular question. Another question was the age of the companies' machinery and their overall level of automation. We also asked separately about the current state of the implementation of Industry 4.0 at the respective companies, in order to understand the situation of these companies at different stages. Due to the complexity of the whole survey, we only asked a simple question about COVID-19: how has the COVID-19 pandemic affected the adoption of Industry 4.0 in your company? There were three possible answers: did it speed up, slow down or have no impact.

The survey focused on the domestic automotive industry and was completed anonymously by 129 representatives of different companies between spring and summer of 2021. Compared to the Hungarian automotive industry as a whole, the responses of these representatives are considered significant. The proportion of domestic and foreign-owned companies completing the questionnaire is illustrated in

Figure 2. In addition, Figure 3 shows the size of the respondents' companies (according to Hungarian rules, a company is considered a large company if it employs more than 250 people). Figure 4 presents the general age of the respondents' machinery (average machinery less than 10 years old is considered modern). Figure 5 indicates the overall level of automation of the respondent's machinery (semi- and fully automated production machines and production lines were counted in the automated category, everything else was included in the non-automated section).

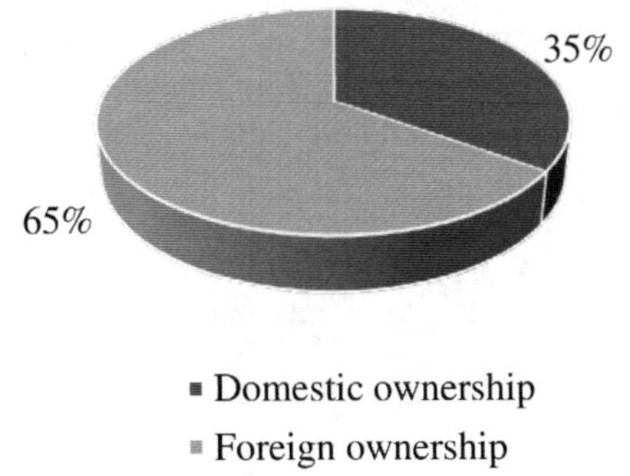

Figure 2. Nationality of owners of companies completing the survey (own figure).

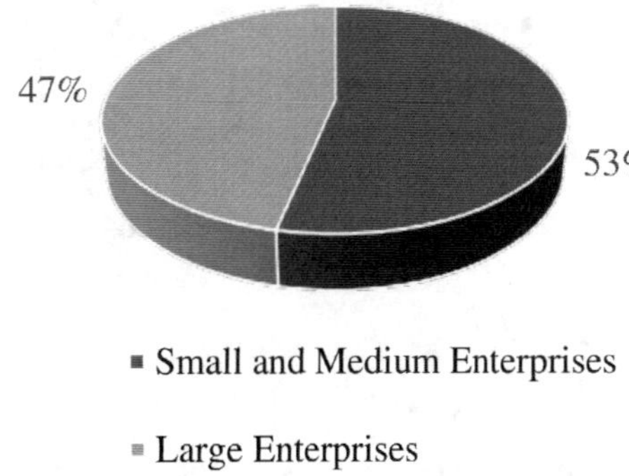

Figure 3. Size of companies completing the survey (own figure).

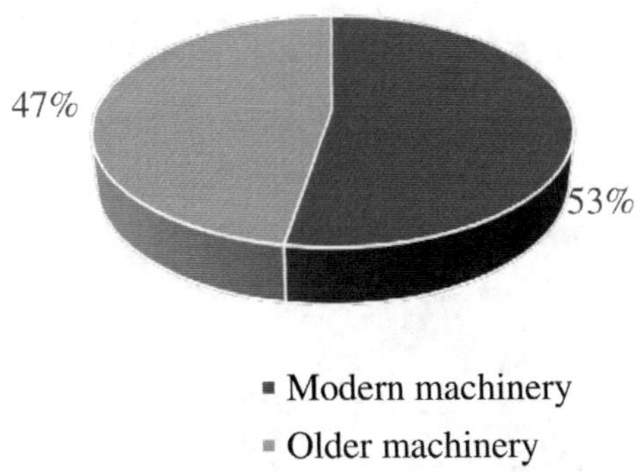

Figure 4. The age of the machinery of the companies completing the survey (own figure).

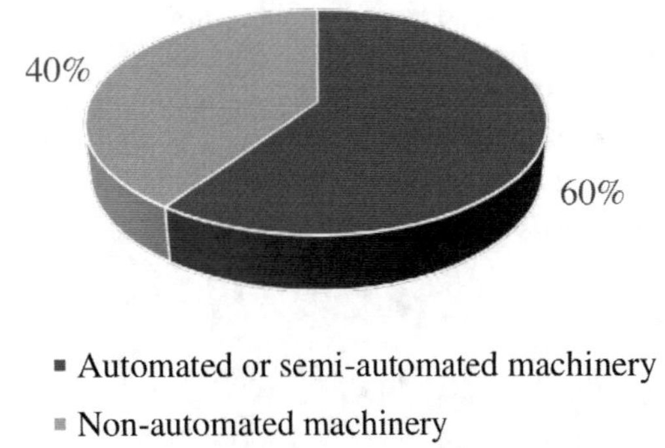

Figure 5. The level of automation of the machinery of the companies completing the survey (own figure).

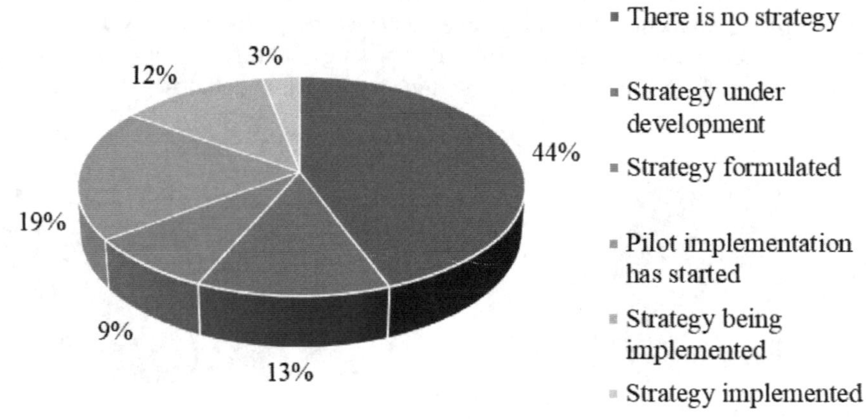

Figure 6. Stage of implementation of Industry 4.0 by the companies completing the survey (own figure).

3. Results and analysis

This chapter analyses the opinions of Hungarian automotive maintenance professionals, separated to different subgroups of respondents mentioned earlier.

Based on the responses, there is no significant difference in the impact of the pandemic on the adoption of Industry 4.0 in terms of company ownership (Figure 7) or company size (Figure 8). Approximately half of the respondents (46-56%) across all types of companies agreed that COVID-19 had accelerated these processes, with the remaining 50% split roughly equally between those who said it had slowed down (21-33%) and those who said it had had no impact (17-27%). This suggests that the changes brought about by the COVID19 pandemic have had a mostly positive impact on Industry 4.0 adoption and industrial digitalization, according to the maintenance departments of domestic manufacturing companies.

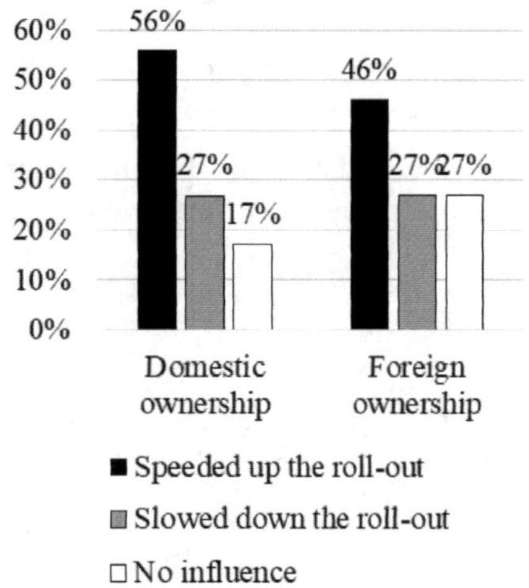

Figure 7. Comparison of the COVID-19 effect by nationality of ownership of companies (own figure).

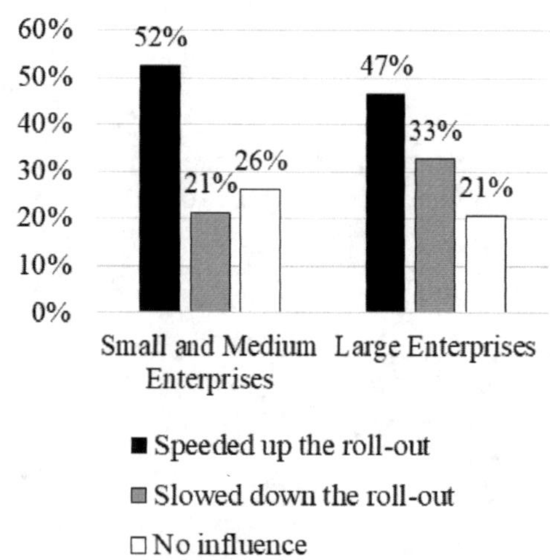

Figure 8. Comparison of the COVID-19 effect by company size (own figure).

In terms of the modernity of the machinery of the companies, similar proportions can be observed (Figure 9). There is, however, a greater difference in the level of automation of the machinery in the companies (Figure 10). Companies with automated and semi-automated machines have a much lower proportion of positive respondents (42%) and a much higher proportion of negative respondents (32%). For companies with non-automated equipment, the trend is the opposite. Here, the proportion of respondents who are in favor of accelerating Industry 4.0 adoption is significantly higher (61%), while the proportion of respondents who are in favor of slowing down adoption and those who are neutral are the same (20-20%). This suggests that the maintenance departments of domestic manufacturing companies believe that the changes brought about by the COVID19 pandemic have had a positive impact on the introduction of Industry 4.0 and industrial digitalization, primarily on less advanced companies with non-automated equipment.

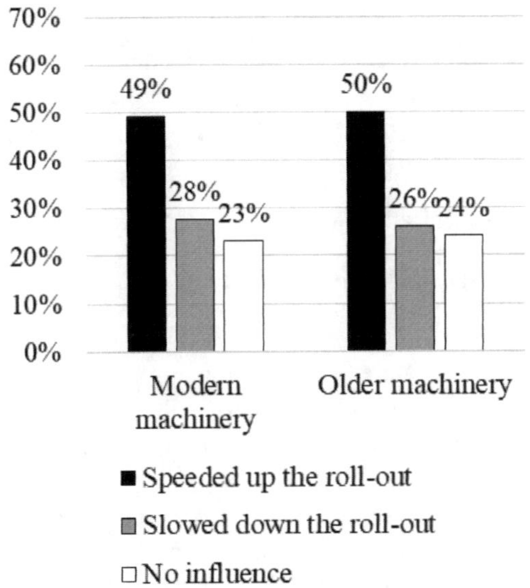

Figure 9. Comparison of the COVID-19 effect with the modernity of companies' machinery (own figure).

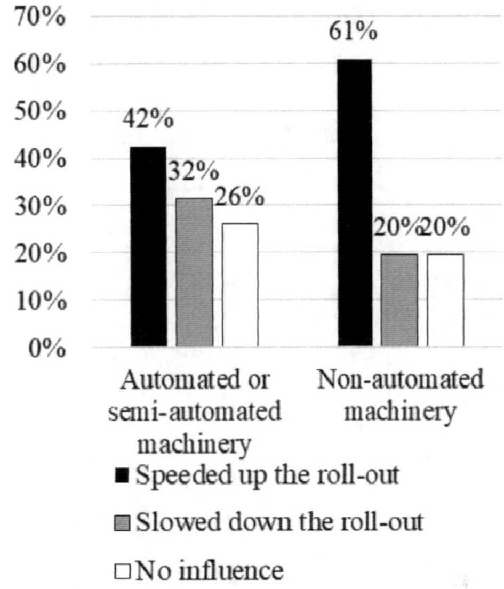

Figure 10. Comparison of the COVID-19 effect with the level of automation of companies' machinery (own figure).

It is also worth looking at how different companies at different stages of Industry 4.0 implementation have been affected by the pandemic (Figure 11). In the early stages of Industry 4.0 adoption (there is no strategy or strategy under development stage), a high acceleration effect can be found (58-60%), similar to the previous figures. Based on our survey, these companies are the most numerous in the Hungarian automotive sector (Figure 6). However, the trend is reversed at the next stage, at companies with a formulated strategy showing a predominance of companies with a strongly negative impact (55%), while those with a positive view have declined to 36%. Thereafter, as the state of implementation continues to evolve, the response rate becomes increasingly positive. At the start of the pilot implementation, all three responses are at about the same level (35-35-30%), and then at full implementation the previously seen 50-25-25% ratio returns.

This shows that the maintenance departments of domestic manufacturing companies consider that the impact of COVID-19 depends on how advanced the company is in implementing Industry 4.0. For the vast majority of the advanced states (this includes the vast majority of companies), the pandemic had a positive impact. However, in some cases the perception is unclear (pilot implementation has started) and in some cases it has explicitly had a negative effect (strategy has been formulated). These are the phases where most of the investment in an Industry 4.0 implementation takes place (e.g. system deployment, training of support staff). These companies support other projects during the pandemic, which can slow down further industrial digitalization.

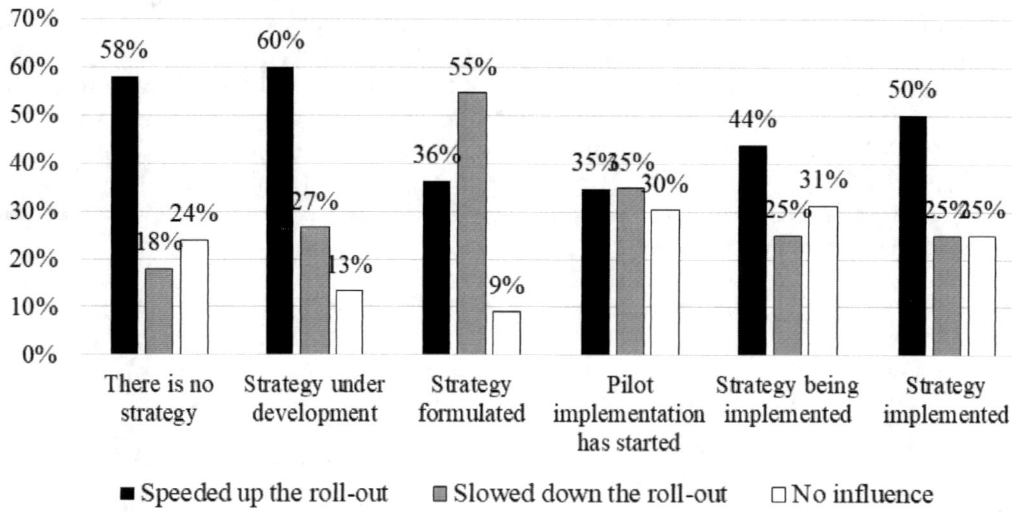

Figure 11. Comparison of the COVID-19 effect with the stage of development of companies in Industry 4.0 (own figure).

4. Conclusions

Based on the opinions of the Hungarian automotive maintenance professionals surveyed, the analysis of the impact of the COVID19 pandemic on the adoption of Industry 4.0 and industrial digitalization led to the following main conclusions:

- There is no significant difference in the impact of pandemic on the adoption of Industry 4.0 across different types of companies, it has had a mostly positive impact on the introduction of Industry 4.0 and industrial digitalization. Approximately half of companies believe that the pandemic has accelerated the adoption of Industry 4.0, thus has boosted industrial digitalization. The remaining 50% is split roughly equally between those who think the pandemic slowed down the adoption and those who think it had no impact.
- There was only one different company type parameter, namely the degree of automation of the machinery. Here the positive impact of the pandemic was particularly high for manufacturing companies with less automated equipment. On the other hand, the companies with the highest level of automation had the lowest positive response rate (still the highest number of answers) in all comparisons.
- The impact of COVID-19 depends on how advanced the company is in implementing Industry 4.0. For the vast majority of the implementing stages (including the vast majority of companies), the pandemic had a positive impact. However, in one case (stage pilot implementation has started) the evaluation of the impact is divided, and in another case (stage strategy has been formulated) it is explicitly negative.

5. Summary

In the Hungarian automotive industry, a huge upsurge in interest in Industry 4.0 and various industrial digitalization initiatives have been observed recently. However, the COVID-19 pandemic has fundamentally changed the budgets, development directions and basic work organization of companies. This is true for all departments of a company, especially industrial maintenance processes, which could be one of the biggest beneficiaries of Industry 4.0. Our paper shows partial results, specific to the

COVID-19 pandemic experiences of a larger, longer-term study, where we measured the overall digitalization of the domestic maintenance sector.

In this paper we described the sampling methodology of our research. Our study is broken down into different types of companies so that we could draw different further conclusions based on several parameters. These parameters included the nationality of the owner, the size of the company, the age and degree of automation of the machinery, and the current state of implementation of Industry 4.0. One conclusion was that approximately half of all companies believe that the pandemic has accelerated the adoption of Industry 4.0, thus boosting industrial digitalization. The remaining 50% is split roughly equally between those who think the pandemic slowed down the adaption and those who think it had no impact.

In the period ahead, we intend to focus on the conclusions in more depth, in the sense that we will propose solutions to the resulting difficulties, according to the type of company. Furthermore, we intend to develop further themes of our final, large survey in a similar way to the present paper.

References

[1] Ľubomír A and Katarína T 2021 Proactive Maintenance as a Tool of Optimization for Vehicle Fleets, in Terms of Economic and Technical Benefits *Acta Polytechnica Hungarica* Vol. 18 No. 8

[2] Bujdosó T 2021 How has the Hungarian and global car industry fared during the epidemic and what can be expected afterwards? https://www.eulerhermes.com/hu_HU/hirek-es-elemzesek/friss-hirek/hogyan-alakult-a-magyar-es-globalis-autoipar-a-jarvany-alatt-es-mi-varhato-utana.html (the last downloading: 28.12.2021)

[3] Khamis A Meng J Wang J Azar AT Prestes E Li H Hameed IA Takács Á Rudas IJ Haidegger T 2021 Robotics and Intelligent Systems Against a Pandemic *Acta Polytechnica Hungarica* Vol. 18 No. 5

[4] Uetz AM and Ellis NJ 2021 Continued Disruption in the Global Supply Chain - and Strained Balance Sheets - Threatens An Already Challenged Automotive Supply Industry https://www.natlawreview.com/article/continued-disruption-global-supply-chain-and-strained-balance-sheets-threatens (the last downloading: 28.12.2021)

[5] Túry G 2021 The transformation of the car industry is inevitable - mass redundancies are no longer an option https://www.portfolio.hu/krtk/20210514/elkerulhetetlen-az-autoipar-atalakulasa-tomeges-leepitesekkel-ez-mar-nem-uszhato-meg-483062 (the last downloading: 28.12.2021)

[6] Durucu M and Bayraktar CA 2020 The Effect of Working from Home on Work and Private Life: Automotive Sector *Application Industrial Engineering in the Digital Disruption Era* pp 148-155

[7] Motavalli J 2021 Phasing Out Internal Combustion Engines? It's Already Happening https://www.autoweek.com/news/a36292118/phasing-out-internal-combustion-engines/ (the last downloading: 28.12.2021)

[8] Juhász L and Pokorádi L 2018 Augmented reality in modern maintenance *Repüléstudományi Közlemények* 30 p 37-46 (Szolnok)

[9] Juhász L and Pokorádi L 2018a Interconnections between Internet of Thinks and maintenance in our days *GRADUS* 5 p 99-106. (Kecskemét)

[10] Juhász L and Pokorádi L 2021 Industry 4.0 and Modern Maintenance in Today's Hungarian Vehicle Industry *J. Phys.: Conf.* Ser. 1935 01 2008

[11] Nick GA 2018 Challenges of adopting the Industry 4.0 paradigm in Hungary in the context of its territorial and company aspects (Győr: Széchenyi István University)

IOP Conf. Series: Materials Science and Engineering 1237 (2022) 012006 doi:10.1088/1757-899X/1237/1/012006

An Improved Equipment for Measuring Heating Caused by Reactive Loads

Dr. János Bihari[1], Imre Marada[2],

[1]Associate Professor, University of Miskolc, Institute of Machine and Product Design, 3515 Miskolc, Miskolc-Egyetemváros, machbj@uni-miskolc.hu

[2]PhD Student, University of Miskolc, Institute of Machine and Product Design, 3515 Miskolc, Miskolc-Egyetemváros, maradaimre@gmail.com

Abstract. In the case of gears that also perform a support function, taking into account the effect of reactive loads is an essential element of sizing. Pulsating or continuous but variable reactive loads can cause heating similar to or greater than in case of normal operation. This heating can be such that it significantly affects both the operating clearance and the strength characteristics of the gears. The effects of reactive loads were previously measured with the help of special drive units, which proved to be a useful method. However, due to the design of the drive units, these measurements were only able to demonstrate that the effect of the reactive loads is significant for the sizing even if they are not of a magnitude causing immediate failure of the gears. Our new research aims to quantify these effects for as many applications as possible. In this article, we introduce the features of the old method and the solutions, which aim to eliminate their disadvantages.

1. Definitions

Small gear: In our work, "small gear" is a gear with a module not larger than 0,5 mm and a maximum characteristic size under 30 mm.

Plastic gear: In our work, we applied cylindrical gears with involute profile and external straight teeth, made of polyamides or polyoxymethylene.

2. The drive units used previously

Structures that fix the gears in the correct positions and allow for drive and load are required for in-service testing. With these it is possible to test the entire system, e.g., various bearings. For this purpose, we typically use test benches, which are usually very expensive for small plastic gears. For this reason, we have formerly designed a special type of drive unit for the tests of the recirculation torque and the reactive loads. When this was placed into simple devices, we were able to perform tests with it at specific settings, and we could eliminate the difficult adjustment possibilities of the expensive test benches. It was taken into account during the creation of this special equipment, that in several cases during the analysis of small plastic gears, it is usually easier to rotate the entire drive unit, because this makes it possible to install static torque measuring equipment. These drive units, that were used for the previous tests, were created with the consideration of the following basic principles [5]:

1. The drive units need to contain a pair of gears.
2. We need to be capable of the measurement of the recirculation torque and the reactive loads. We also need to be able to combine these tests when it is necessary.

3. We must ensure, that there can be no collision problems with the clamps, thus they need to have rounded outer corners.
4. It is important, that the rigidity must be high in bearings.
5. For the gears we need to create enough space, the various possible layouts and the wheelbase tolerances must be taken into account.
6. A direction perpendicular to the axes of the gears is required for the division of the housings, for mounting purposes. We must also ensure that the positioning of the housing halves is accurate enough.
7. We need to reduce the weight of the housings to be as small as possible. If they are heavier, they can have a negative effect on the recirculation torque tests.
8. We must produce them in as many variations as possible with the tools at our disposal.

The last point of the requirement list in our case meant a Roland Modela CNC milling machine, which can produce the parts for the drive unit with acceptable precision. However, it would have taken 60 hours to manufacture a single housing if we wanted to ensure that point 7 of the requirement list was not ignored. Thus, we manufactured the drive units with FDM 3D printing from ABS material and finishing was made with precision milling. We also had a problem with determining the cutting parameters for the precision milling, so we conducted experiments, as we had no literature or experience at our disposal about the machining of 3D printed parts. These experiments were done on many different materials, we obtained the perfect results with the ABS material, so we chose that. With this method we were able to limit the expensive and slow precision milling to the finishing of the holes of the shafts and bearings. This way we reduced the manufacturing time of the drive units to a mean value of two hours [5].

Essentially, two types of drive units were developed and manufactured in several versions, one for the use of plain bearings (Figure 1) and another for the use of roller bearings (Figure 2). At the time these drive units were created, the tests required four different wheelbases and another design, which simulated the typical faults due to housing distortions, so five versions were made. We designed these different wheelbases in accord with the most important ISO tolerances. The error that was simulated was the skew which is formed when the housing of the drive unit is distorted. For this we inclined the bearing of a shaft by 3° [1] [5].

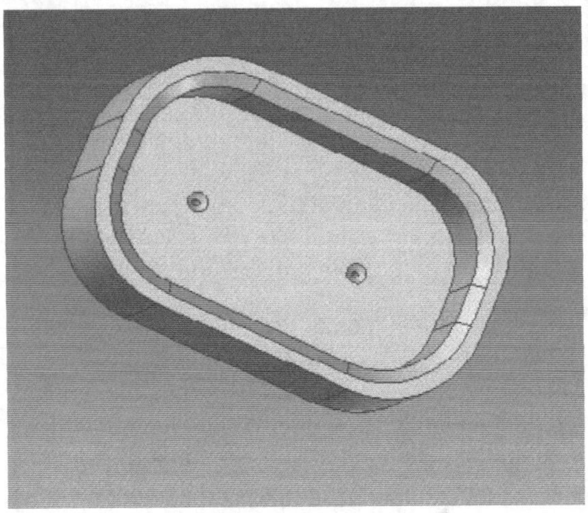

Figure 1. The housing created for plain bearings [5]

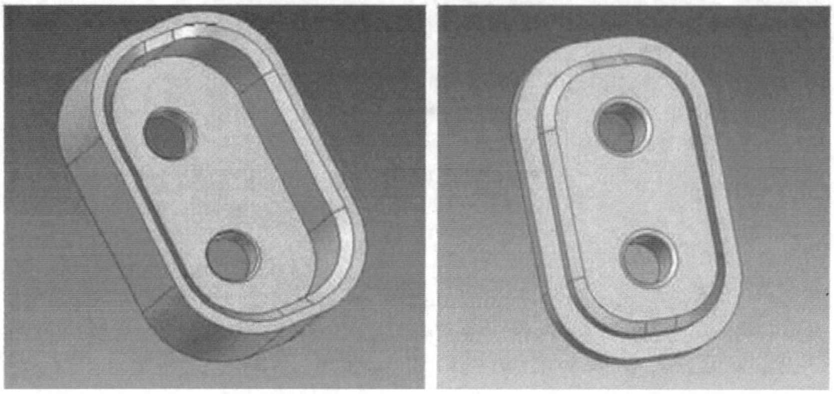

Figure 2. The housing created for the use of roller bearings [5]

3. Usance of the former drive units and the accomplished results

3.1. Usance of the equipment in the reactive load tests

Difficult-to-diagnose problems can be generated by the effects of reactive loads in drive units fitted with small plastic gears especially when they have a support function too [4]. Since this function is a commonly used additional utilization, principally for drive units with more than one stages, the thorough research of the effects of reactive loads is highly needed. To analyse these effects, we fixed one of the gears and used a repetitive load on the shaft of the other gear. At the time, we fixed the gear with a screw through a hole, and we applied the load through a steel weight. Figure 3 shows the layout of the reactive load test. We attached a lever (3) straight to the output shaft. The drive unit (1) and the drive motor (5) were mounted on a base (2). An eccentric disc (4) was situated on the shaft of the drive motor. The eccentric disk was under the lever, so when the motor was operating, the eccentric disk raised the lever, and the lever loaded the gears. We created this load with the help of a weight (6), which was suspended by a joint (7) at the end of the lever. When we adjusted the location of the drive motor, the angular rotation of the lever was also adjusted. This way, the number of loads corresponds to one per the rotation of the motor, so the speed of the motor determines the number of loads per time duration [5].

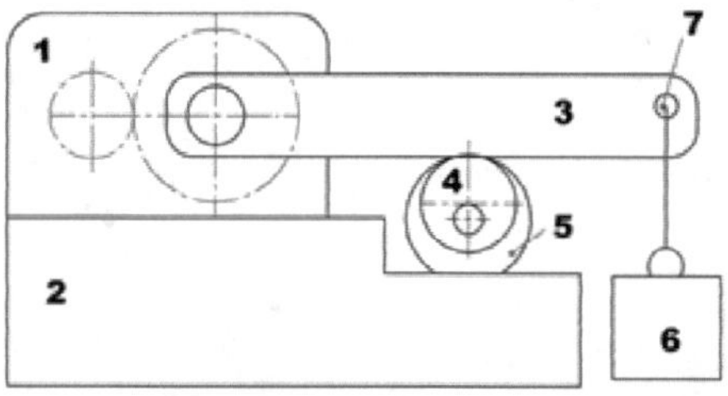

Figure 3. Layout of the test [5]

3.2. Results of the reactive load tests

We made several tests with the previously used drive units. These tests showed, among other things, that considerable and dangerous local temperature rises can be generated by the effects of reactive loads in the connected teeth [5].

4. The reasons that lead to the development of new drive units

With the drive units used so far, the effects of the reactive loads can be demonstrated, but they are no longer suitable for determining the exact values of the effects. A major problem is that measurements can only be made on a single stage, which is disadvantageous in several respects.

In gearboxes that also have a support function, the drive motor's usually small moment of inertia is very high, and typically acts on the supported element through a ratio of between 1:400 and 1:1600. However, this means a large number of associated components, as well as backlash at each connection. Therefore, testing with one stage and one of the gears locked is not accurate because only the number of teeth corresponding to the contact ratio is connected, whereas in real cases there is always a displacement due to backlash in the system. This displacement can be beneficial for heating, especially if the held element flexibly returns to its original position, such as air dampers with hollow seals, because more teeth are engaged than in the fixed case. However, in the fixed case, there is typically no wear, which is also not true here, because many of these drive units have been shown to experience wear on the tooth surfaces due to small but high repetition rate displacements.

Another characteristic of such plastic drive units is that, due to the large number of stages, the speed and torque differences between the input and output of the gearbox are also very large. To minimise the volume, it is therefore common to vary the modules, number of teeth and tooth widths for the same gear ratio, even per stage. For this reason, the effect of reactive loads does not necessarily only occur on the last stage.

In addition, the fixed solution does not allow the positions to be changed during the test. Damage from repetitive reaction loads occurs when the drive unit starts to operate after a prolonged period of inactivity.

The experiments so far have been carried out at ambient temperatures between 20 and 25 °C. However, the effects of reactive loads are always more dangerous at higher ambient temperatures, so measurements at ambient temperatures of at least 60 °C are also necessary. A further disadvantage of old drive units is that their housing is plastic. This is a common solution in real applications, but due to the thermal expansion of the housings, the wheelbases and thus the operating clearances change (by up to an order of magnitude) more with ambient temperature than with steel housings. Thus, the tests have to deal with both the change in the wheelbases and the change in the gears' own dimensions. As we are currently mainly interested in the behaviour of the gears, this complicates the evaluation of the results considerably.

Typically, drive units that also have a supporting function must be designed so that if the reactive load exceeds the designed supporting force, e.g., in the case of mirror drive units in cars, when the mirror plate is pressed, the drive unit can rotate backwards without damage. In such cases, the drive unit may be subjected to several times the designed working load. Such a test cannot be carried out with fixed gears.

5. Requirements of the new drive unit

With the formerly used drive units we could not create different gear ratios, i.e., we were able to install only gears with equal number of teeth. With the enhanced new drive unit, we must solve this problem. The solution can be done with eccentrics or replaceable covers [5].

The surfaces of the old drive unit used for fastening did not make it possible to return the equipment to the clamp between tests in a changeless way within 0,05 mm in the x and y directions, which created an altering impact during rotation. We need to make possible the accurate repetition with the new equipment [5].

However, while the old drive units were required to be minimal weight so that their mass did not affect the analysis of the recirculation torque, this is no longer a relevant consideration for the new ones. If the drive unit fits exactly into the required devices, the effect of the higher mass can be determined by simple calibration. This allows the drive unit housings to be made of steel, which is important for analysing the effect of reactive loads.

In our previous tests we were not able to combine the old drive units, so we could not test multi-stage drives. At the time the combination of drive units was not necessary, because the gear ratio was always 1:1. However, in our future research we would like to study the interaction between the faults of the gears, so it has a bigger importance to be able to create several stages in our upgraded equipment [5].

When we created the old equipment, the bearing bush of the housing with the plain bearings was manufactured of the plastic material, which was used for the housing. Manufacturers already know and utilize injection moulding techniques, which make it possible to combine different polymers. Thus, we must create housings and covers, where the material of the plain bearings and the material of the housing are not the same [5].

Additionally, we still have to take into account many points of the requirements list, mentioned in section 2, these points are the following [5]:

1. We still need to be able to measure the recirculation torque and the reactive loads. We also need to be able to combine these tests when it is necessary.
2. It is still very important, that the rigidity must be high in bearings.
3. For the gears we need still to create enough space. The possible layouts and the wheelbase tolerances must be taken into account.
4. A direction perpendicular to the axes of the gears is still required for the division of the housings, for mounting purposes. But it is recommended, to have multiple divisions, for example a centre part with covers.
5. We still must produce them in as many variations as possible with the tools at our disposal.

6. New drive units for new tests

The housing of the new drive units consists of three parts, a centre housing part and two covers. The precise positioning of the covers is ensured by two positioning pins. The covers are fixed with through bolts.

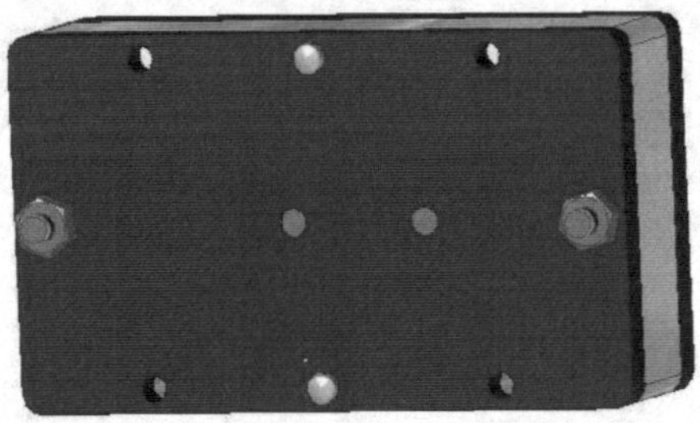

Figure 4. The model of the drive unit

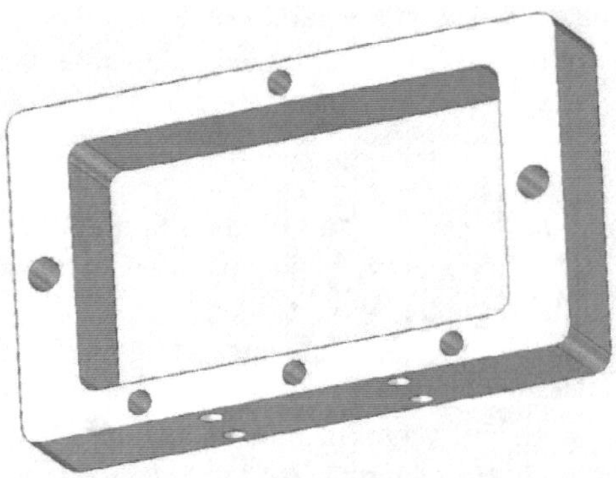

Figure 5. The centre housing part

The centre housing is constant, the same for all ratios and wheelbases. The covers are interchangeable and can be produced in several variations, allowing the design of drives with different wheelbases and ratios. Different wheelbases are associated with different pairs of covers of different designs, of which the appropriate one for the test must be fixed to the housing. Thus, these drive units can contain not only gear pairs with the same number of teeth, but also different ones. These elements are made of steel by electrical discharge machining.

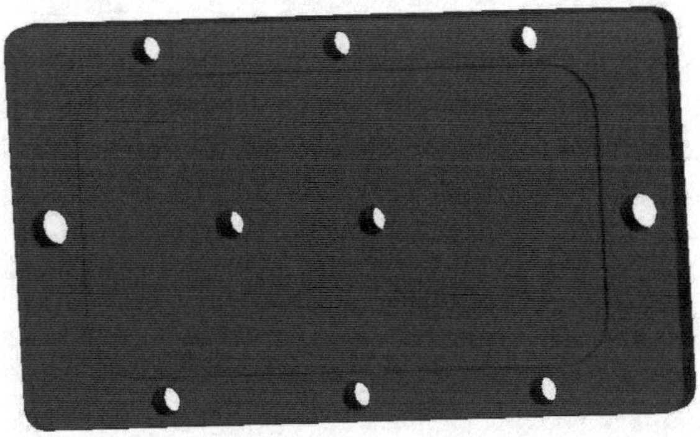

Figure 6. Design of the cover for plain bearings

For the time being, the covers will be made for plain and roller bearings, but other designs are also possible, such as one where plain bearings other than the housing material are used.

Figure 7. Design of the cover for roller bearings

The exact positioning of the drive unit in relation to the clamp is achieved by means of the positioning holes in the centre housing. This means that they always engage the same way in the clamp within the required accuracy, thus allowing accurate repeatability, i.e., a drive unit can be repositioned in the clamp within 0,02 mm in the x and y directions. This is currently sufficient accuracy for all planned tests. In addition, the pin mounting allows for accurate calibration, so that inertia due to the higher mass of the gearboxes is not a problem for other tests.

Figure 8. Positioning holes at the bottom of the drive unit

Figure 9. The inside of the drive unit

It is also possible to combine the new drive units, so we can even create multi-stage drives. This is an important aspect for them, as the gear ratio is no longer 1:1, but can be any desired, depending on the holes in the covers. This makes it possible to investigate the reactive loads and the interaction of faults by designing multiple stages.

Figure 10. The model of three-stage combined drive unit

Figure 11. The combined drive unit

The red elements in Figure 11. are intermediate pieces that provide space for the couplings.

Summary

The new type of drive units allows tests to be carried out in closer-to-reality situations, for example at high ambient temperatures. The steel housings simplify the positioning of the sensors, as they can be provided with holes, so that measurements can be made separately for each step, speeding up data collection. The modular design and the calibratability of the plain bearings also allow to model bearing arrangements with different fits per step, which is common for this type of drive units.

Literature

[1] VDI 2731 Mikrogetriebe, Grundlagen. BeuthVerlag, Berlin, 2009.

[2] Bihari, J.: Kisméretű műanyag fogaskerekes hajtások hibái, *GÉP 65*:(2), pp: 19 – 22, 2014.

[3] Bihari, J. – Kamondi, L.: Kis méretű műanyag fogaskerekek vizsgálata, *GÉP 62*:(7-8), pp: 21 – 24, 2011.

[4] Sarka, F.; Bihari, J.; Takács, Á.** ; Tóbis, Zs.: Test Method for Investigation of Reactive Loads on Gear Drives with Supporting Function, LECTURE NOTES IN MECHANICAL ENGINEERING 22 pp. 265-272., 8 p. (2021) VDI 2736 Thermoplastische Zahnräder, Blatt 2, BeuthVerlag, Berlin, 2014.

[5] Marada, I., Bihari, J.: Designing a new type of drive unit for the analysis of small plastic gears, MULTIDISZCIPLINÁRIS TUDOMÁNYOK: A MISKOLCI EGYETEM KÖZLEMÉNYE 11: 5 pp. 245-250., 6 p. (2021)

IOP Conf. Series: Materials Science and Engineering 1237 (2022) 012007 doi:10.1088/1757-899X/1237/1/012007

Comparison of Methods for Calculating the Heating of Plastic Gears

Imre Marada[1], Dr. János Bihari[2]

[1]PhD Student, University of Miskolc, Institute of Machine and Product Design, 3515 Miskolc, Miskolc-Egyetemváros, maradaimre@gmail.com

[2]Associate Professor, University of Miskolc, Institute of Machine and Product Design, 3515 Miskolc, Miskolc-Egyetemváros, machbj@uni-miskolc.hu

Abstract. Accurate determination of heating is significantly more important for plastic gears, even in simpler applications than for steel gears. These gears are often used without lubrication or with a single, lifetime lubrication. The load capacity of materials is significantly affected by temperature, even in ordinary cases. Even today, heating is often calculated for different heating cases according to the calculation method of VDI 2545 or the modified VDI 2545, although these methods have been proven to be obsolete. Since the main direction of our research is to explore the design features of drive units equipped with small plastic gears, we tried to assess in advance, using a series of calculations, how much difference there is between the results obtained with obsolete and modern calculation methods. In this article, we present the basic characteristics of the calculation methods and compare the results obtained for some m>1 mm and m<0,5 mm gears.

1. Introduction

Heating is a fundamentally bigger problem for plastic gears than for steel gears. The strength of plastics can vary significantly even at temperatures that occur in the everyday environment, such as in places exposed to direct sunlight. In gear drives, this can have a significant effect on the behavior of the drive unit without rapid failure of the components, with higher temperatures leading to greater deformation of the tooth surface, greater wear and consequently a reduction in the contact ratio. At the same time, drive units with plastic gears are often used without lubrication, and temperature also has a strong influence on the tribological properties of plastics. [1]

There are no thorough and detailed standards for the design of plastic gears, such as DIN 3990. A common method is to design such gears according to a common standard for metal gears, simply substituting the strength properties of plastics in the calculations. This method, however, does not address the problem that the strength of plastics can vary significantly with changes in operating ambient temperature. This is addressed by current standards and guidelines for the design of plastic gears, either by a computational method or by a combination of experiments and calculations. In Europe, the VDI 2545 guideline has long been the basis for the design of plastic gears. It was published in 1981 and withdrawn in 1996. Since no other similar guideline was available after its withdrawal, it was generally used in practice until the publication of VDI 2736 in 2016 and is still relevant today. Therefore, in this article we present the calculation method to calculate the heating of plastic gears according to VDI 2545. In addition, the calculation method of VDI 2736 is presented. AGMA and JIS standards for gear design recommend experimental methods for the determination of heating, of which we present here the method

IOP Conf. Series: Materials Science and Engineering 1237 (2022) 012007 doi:10.1088/1757-899X/1237/1/012007

of JIS B 1759. Since there was no generally accepted method of measurement for 10 years, several material manufacturers have developed documents that are well suited to use within given limits. Here we present the method of Licharz company, which is a modified version of VDI 2545 based on empirical data.

2. Calculation methods of the VDI guidelines

2.1. VDI 2545 [2]

$$\vartheta_{1,2} = \vartheta_k + 136 \cdot P \cdot \mu \cdot \frac{1+u}{z_2+5} \cdot \left(\frac{17100}{b \cdot z_{1,2}} \cdot \frac{k_2}{(v \cdot m)^\kappa} + 6{,}3 \cdot \frac{k_3}{A} \right) \quad (1)$$

Where:

- $\vartheta_{1,2}$, the temperature of the gear, °C
- ϑ_K, the ambient temperature, °C
- P, power, kW
- b, width of the tooth face, mm
- μ, coefficient of friction
- z, number of teeth
- m, a module, mm
- v, tangential velocity, m/s
- A, surface of the gear casing, m^2
- u, gear ratio
- k_2, material-related factor
- k_3, gear-related factor, m^2K/W

Table 1. The coefficient of friction and k_3 according to [2]

<table>
<tr><td colspan="2">Lubrication</td><td>Oil mist lubrication</td><td>Assembly lubrication</td><td>Permanent lubrication (grease)</td><td>Dry</td><td>Oil lubrication</td></tr>
<tr><td rowspan="4">Coefficient of friction, μ</td><td>PA-PA,</td><td rowspan="4">0,07</td><td rowspan="4">0,09</td><td rowspan="4">0,04</td><td rowspan="2">0,2</td><td rowspan="4">0,04</td></tr>
<tr><td>PA-steel,</td></tr>
<tr><td>POM-POM,</td><td>0,28</td></tr>
<tr><td>POM-steel</td><td>0,2</td></tr>
<tr><td rowspan="3">k_3, m^2K/W</td><td>open gear</td><td colspan="5">0</td></tr>
<tr><td>partially open gear</td><td colspan="5">0,04…0,13</td></tr>
<tr><td>closed gear</td><td colspan="5">0,17</td></tr>
</table>

Table 2. The value of k_2 and κ according to [2]

<table>
<tr><td colspan="2">Material</td><td>PA-PA</td><td>PA-steel</td><td>POM-POM</td><td>POM-steel</td><td rowspan="3">for immersion and syringe lubrication: 0</td></tr>
<tr><td rowspan="2">k_2</td><td>flank temperature</td><td>15</td><td>10</td><td>7</td><td rowspan="2"></td></tr>
<tr><td>root temperature</td><td>2,4</td><td>1,0</td><td>2,5</td></tr>
<tr><td colspan="2">κ</td><td>0,75</td><td>0,75</td><td>0,4</td><td>0,4</td><td></td></tr>
</table>

2.2. Modified VDI 2545 [3]

$$\vartheta_{1,2} = \vartheta_k + 136 \cdot P \cdot \mu \cdot \frac{1+i}{z_2 + 5 \cdot i} \cdot \left(\frac{17100}{b \cdot z_{1,2}} \cdot \frac{k_2}{(v \cdot m)^{\frac{3}{4}}} + 7{,}33 \cdot \frac{k_3}{A} \right) \quad (2)$$

Where:

- $\vartheta_{1,2}$, the temperature of the gear, °C
- ϑ_K, the ambient temperature, °C
- P, power, kW
- b, width of the tooth face, mm
- μ, coefficient of friction
- z, number of teeth
- m, a module, mm
- v, tangential velocity, m/s
- A, surface of the gear casing, m^2
- i, transmission ratio
- k_2, material-related factor
- k_3, gear-related factor, m^2K/W

Table 3. Value of the factor k_2 according to [3]

For the calculation of flank temperature	For the calculation of root temperature:
k_2=7, for mating components steel/plastic	k_2=1, for mating components steel/plastic
k_2=10, for mating components plastic/plastic	k_2=2,4, for mating components plastic/plastic
k_2=0, in the case of oil lubrication	k_2=0, in the case of oil lubrication
k_2=0, at a velocity not bigger than 1 m/s	k_2=0, at a velocity not bigger than 1 m/s

Table 4. Value of the factor k_3 according to [3]

k_3=0, for completely open gear
k_3=0,043-0,129 m^2K/W, for partially open gear
k_3=0,172 m^2K/W, for closed gear

Table 5. The coefficient of friction according to [3] and [4]

μ=0,04, for gears with permanent lubrication	μ=0,4, PA/PA, dry
μ=0,07, for gears with oil mist lubrication	μ=0,18…0,25, PA/POM, dry
μ=0,09, for gears with assembly lubrication	μ=0,18…0,2, POM/steel, dry
μ=0,2, PA/steel, dry	μ=0,2…0,28, POM/POM, dry
μ=0,35, PA/PBT, dry	μ=0,18, POM/PBT, dry

In the case of intermittent operation, the relative duty cycle ED is defined as the ratio between the load duration t and the overall cycle time T as a percentage [3]:

$$ED = \frac{t}{T} \cdot 100 \quad (3)$$

The value of the correction factor f as a function of the relative load cycle can be determined from a diagram. Thus, the equation modified by the correction factor f for intermittent operation [3]:

$$\vartheta_{1,2} = \vartheta_k + 136 \cdot P \cdot \mu \cdot f \cdot \frac{1+i}{z_2 + 5 \cdot i} \cdot \left(\frac{17100}{b \cdot z_{1,2}} \cdot \frac{k_2}{(v \cdot m)^{\frac{3}{4}}} + 7{,}33 \cdot \frac{k_3}{A} \right) \tag{4}$$

2.3. Calculation method of VDI 2736

The tooth root temperature [4]:

$$\vartheta_{ft} = \vartheta_k + H_v \cdot P \cdot \mu \cdot \left(\cdot \frac{k_{\vartheta,Fuß}}{b \cdot z \cdot (v \cdot m_n)^{\frac{3}{4}}} + \frac{R_{\lambda,G}}{A_G} \right) \cdot ED^{0,64} \tag{5}$$

The flank temperature [VDI 04]:

$$\vartheta_{Fla} = \vartheta_k + H_v \cdot P \cdot \mu \cdot \left(\frac{k_{\vartheta,Fla}}{b \cdot z \cdot (v \cdot m_n)^{\frac{3}{4}}} + \frac{R_{\lambda,G}}{A_G} \right) \cdot ED^{0,64} \tag{6}$$

Where [4]:

- ϑ_K, the ambient temperature, °C
- P, nominal output, W
- μ, coefficient of friction
- H_v, degree of tooth loss
- $k_{\vartheta,Fuß}$, heat transfer coefficient of the plastic gear for the calculation of root temperature
- $k_{\vartheta,Fla}$, heat transfer coefficient of the plastic gear for the calculation of flank temperature
- b, face width, mm
- z, number of teeth
- v, tangential velocity, m/s
- m_n, normal module, mm
- $R_{\lambda,G}$, heat transfer resistance of the mechanism housing
- A_G, heat-dissipating surface of the mechanism housing, m^2
- ED, relative tooth-engagement time

The degree of tooth loss [4]:

$$H_v = \frac{\pi \cdot (u+1)}{z \cdot cos\beta_b} \cdot (1 - \varepsilon_1 - \varepsilon_2 + {\varepsilon_1}^2 + {\varepsilon_2}^2) \tag{7}$$

Where [4]:

- u, gear ratio
- z, number of teeth
- β_b, helix angle at the base circle
- ε_1 and ε_2, partial radial contact ratios of pinion and gear

$$\varepsilon_i = \frac{z_i}{2 \cdot \pi} \cdot \left(\sqrt{\left(\frac{d_{ai}}{d_{bi}}\right)^2 - 1} - tan\alpha_{wt} \right) \tag{8}$$

- d_{ai}, addendum circle diameter, mm
- d_{bi}, base circle diameter, mm
- α_{wt}, operating pressure angle in the transverse section

Table 6. Heat transfer coefficients in $K(m/s)^{0,75}mm^{1,75}/W$ [4]

Lubrication	Pairing	$k_{\vartheta,Fuß}$	$k_{\vartheta,Fla}$

circulating oil	all	0	0
dry/grease/oil mist	plastic/plastic	$2{,}1x10^3$	$9x10^3$
	plastic/steel	$0{,}9x10^3$	$6{,}3x10^3$

Table 7. Heat transfer resistance of the mechanism housing in K·m2/W [4]

Mechanism	$R_{\lambda,G}$
open housing	0
partially open housing	0,015…0,045
closed housing	0,06

3. Comparison of calculation methods in the VDI guidelines

All three VDI guidelines relate ambient temperature to the rest of the equation by addition. This means that none of the methods takes into account the fact that not only the strength but also other properties, such as tribological properties or elasticity, can change significantly with temperature in plastics.

Similarities in the VDI methods are that the power P, friction coefficient μ, width b, number of teeth z, tangential velocity v and module m are considered in a very similar way. Furthermore, all three methods take into account the size of the gear surface.

The VDI 2545 method can only be used for continuously running drive units. However, the modified VDI 2545 and VDI 2736 methods are not only applicable for continuous operation, but also for intermittently operating gear units. The former can be modified by a correction factor f, which is a function of the relative load cycle ED, the latter introduces this ED value directly into the formula. This is advantageous, because plastic drive units often only perform setting tasks, i.e., they operate intermittently.

VDI 2545 and the modified VDI 2545 show significant similarities in structure and parameters. The differences are the empirical quantity of 6,3 in the former formula and 7,33 in the latter, and the possible values of the factors k_2 and k_3. VDI 2736 differs from the other two VDI methods in a number of respects, both in its structure and in the parameters it uses.

The material properties are taken into account by VDI 2545 and the modified VDI 2545 with the factor k_2. This factor can also take a different value when calculating the temperature of the tooth flank or the tooth root. For VDI 2736, there are two separate factors for calculating the temperature of the tooth root and the tooth flank. The factor $k_{9Fuß}$ is the factor for the root and k_{9Fla} is the factor for the surface temperature. Both factors depend on the gear material.

The VDI 2545 method includes a factor that the other two VDI methods do not. This is the κ factor, which depends on the gear material pairing. This is one of the reasons why this guideline is still often used today, because plastic gears are often paired with steel gears or plastic gears with different properties, for example to reduce noise or operating clearance. The modified VDI 2545 and VDI 2736 take this κ factor as 0,75, although according to Miklós Antal [2] κ is not 0,75 for all materials.

Various factors are also taken into account when determining the closure of the gearbox. VDI 2545 and modified VDI 2545 use the factor k_3 for this purpose, VDI 2736 uses $R_{\lambda,G}$. They have in common that three different values can be taken depending on whether the drive unit is open, partially closed or closed.

4. The JIS B 1759 standard

The Japanese standard JIS B 1759 presents a possible definition of the strength of plastic gears. For this purpose, the standard describes a test bench, which is used by taking measurements with the standard test parameters presented in the standard, as well as with different parameters, and then comparing them.

The JIS B 1759 standard considers the effect of temperature on the allowable dedendum bending stress of the gear material [5]:

$$\sigma_{FP} = \sigma_{Flim} \cdot Y_{NT} \cdot Y_{\theta} \cdot Y_{\Delta\theta} \cdot Y_{l} \cdot Y_{M} \quad (9)$$

Where [5]:

- σ_{Flim}, the allowable bending stress of gear material, which can be determined by tests
- Y_{NT}, the life coefficient
- Y_M, mating gear coefficient
- Y_L, lubrication coefficient
- Y_θ, ambient temperature coefficient
- $Y_{\Delta\theta}$, temperature increase coefficient

The ambient temperature coefficient for standard test parameters, i.e., for an ambient temperature of 23°C is $Y_\theta=1$. The ambient temperature coefficient for a given temperature other than the standard test conditions is calculated by the following method [5]:

$$Y_\theta = \frac{\sigma_{Flim(at\ \theta=\theta^*)}}{\sigma_{Flim}} \quad (10)$$

Where σ_{Flim} is the allowable bending stress of the gear material at the standard test parameters (ϑ_K=23°C), $\sigma_{Flim(at\ \theta=\theta^*)}$ is the allowable bending stress of the gear material at an ambient temperature different from the standard test parameters, the value of which can be determined by test bench tests. [5]

The temperature increase coefficient is a factor that takes into account the effect of the temperature increases due to friction and hysteresis on the allowable bending stress of the gear material. The temperature growth factor can be obtained as follows [5]:

- If the gear module or tooth width is smaller than the standard test gear: $Y_{\Delta\theta}>1$
- If the gear module or tooth width is greater than the standard test gear: $Y_{\Delta\theta}<1$
- If the rotation rate of the gear is less than the standard test gear: $Y_{\Delta\theta}>1$
- If the rotation rate of the gear is greater than for the standard tests: $Y_{\Delta\theta}<1$

Thus, the methodology of JIS B 1759 is very different from that used in the VDI guidelines. On the one hand, JIS B 1759 does not derive much of the factor from tables and diagrams of previous data but describes how to make measurements to obtain factors that affect strength.

In addition, JIS B 1759 does not specifically address the calculation of the gear temperature. Unlike the VDI guidelines, it incorporates the effect of temperature in the form of factors into the strength calculation formula.

Another important difference is that JIS B 1759 does not neglect the effect of ambient temperature on the gear material but presents a possibility to determine the effect of ambient temperature as a factor in the form of separate measurements.

In addition, however, it does not use a specific relationship to determine the additional temperature increase but presents a benchmark that can be obtained by comparing the parameters of the drive unit we use with the standard test parameters. This method carries some uncertainty, as the standard does not specify the scale at which a change in each parameter will change the factor, only whether it will be greater or less than one.

The temperature increase coefficient does not take into account the tribological properties of the gear material, nor the friction, nor the degree to which the gear is closed or open. It only takes into account the variation of certain geometrical dimensions of the gear and the speed. Furthermore, the JIS B 1759 standard cannot take into account intermittent operation.

5. Other calculation methods

Gear temperature according to [6]:

$$\vartheta = \vartheta_K + \Delta\vartheta_b + \vartheta_f \quad (11)$$

Where ϑ_K is the ambient temperature, the increase in gear body temperature is [6]:

(12)

$$\Delta\vartheta_b = \frac{0{,}625 \cdot \mu \cdot T}{C_p \cdot \rho \cdot z \cdot L \cdot (r_0{}^2 - r_p{}^2)}$$

Where [6]:

- C_p, the heat capacity of air
- ρ, the air density
- T, the transmitted torque
- L, the gear face width
- r_0, the outer radius of the gears
- r_p, the pitch radius

The flash temperature [6]:

$$\vartheta_f = \frac{0{,}555 \cdot \delta H}{L \cdot (k \cdot \rho \cdot c \cdot w \cdot v_s)^{-1/2}} \quad (13)$$

Where [6]:

- the instantaneous energy loss due to friction

$$\delta H = \frac{\pi \cdot P \cdot \mu}{z} \quad (14)$$

- k, the thermal conductivity
- ρ, the density
- c, the specific heat of the surfaces
- v_s, sliding velocity
- w, the Hertzian contact width
- P, the transmitted power
- z the number of teeth

The formula, similar to those in the VDI guidelines, does not take into account the effect of ambient temperature on the gear material. The difference is that the formula takes completely different parameters into account to calculate the temperature increase.

Compared to other temperature formulas, it breaks down the temperature change into two separate terms: one for the temperature inside the gear body and one for the temperature on the surface. The former depends on the torque and gear dimensions as well as on the air, while the latter depends on material properties, surface friction and Hertzian contact.

6. Calculations based on the methods presented

6.1. Definitions

A relevant temperature difference in this interpretation is a difference that exceeds 5 percent of the initial value.

Small gear: a gear with a maximum module of 0,5 mm and a maximum characteristic size of less than 30 mm.

6.2. Calculations

Using the methods presented here, calculations have been carried out for a number of different modules, ratios and powers, for a number of different materials. We were interested to see if there were any relevant differences between the results and, if so, to what extent the size of the gears affected the results.

For the three VDI procedures, the only relevant differences in the module range above 0,5 mm are in extreme cases where the applicability of a particular material or size range is questionable anyway. However, for small gears, relevant differences can be detected, especially for module of 0,3 mm and below. Table 8 is an extract of the calculations. These calculations are made here for POM material with 10/50 gear ratio and 0,3 mm module.

For this gear pair, the tooth surface load capacity at the critical starting temperature of 60 °C, which is not an exceptional value for example in a car dashboard in summer, at 10 watts of power and 10,000 rpm, the tooth surface safety factor is only 1,8, which is not a practical value for continuous operation, but is acceptable for intermittent operation.

Table 8. The results of the calculation

VDI 2545					Modified VDI 2545				VDI 2736			
Assembly lubrication			Dry		Assembly lubrication		Dry		Assembly lubrication		Dry	
	Flank	Root	Flank	Root	Flank	Root	Flank	Root	Flank	Root	Flank	Root
10 W	62,247	61,18	66,99	63,66	63,06	61,255	66,81	62,79	60,041	60,016	60,09	60,04
5 W	61,123	60,59	63,49	61,83	61,53	60,628	63,40	61,39	60,020	60,008	60,05	60,02

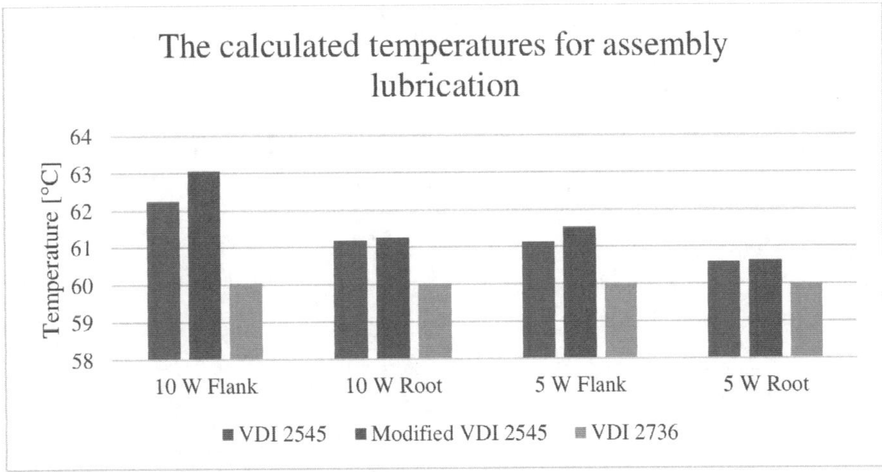

Figure 1. Diagram of the calculated temperatures for assembly lubrication

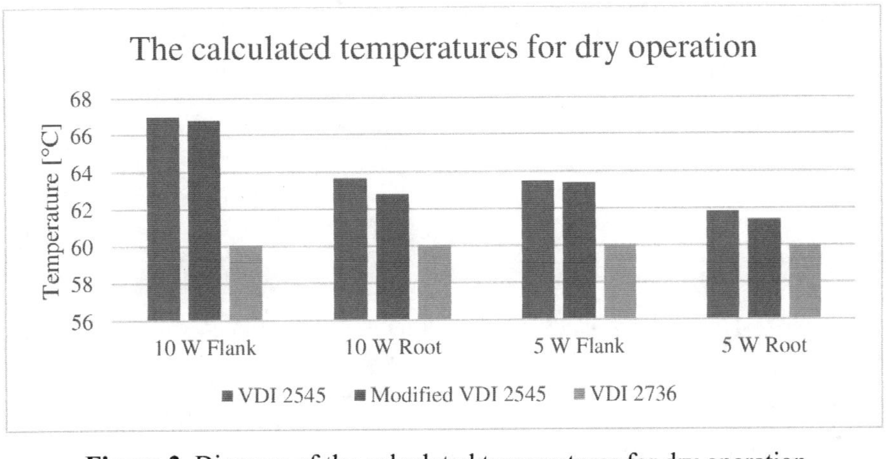

Figure 2. Diagram of the calculated temperatures for dry operation

The values in the table show that VDI 2545 and VDI 2736 already give relevantly different values here. Calculations according to JIS require experiments to be carried out and are therefore not yet included in this calculation.

Summary

This article presents five methods that can be used to determine the in-service heating of plastic gears. Three of these methods have been used for comparative calculations to determine the extent to which the results obtained by the different methods differ from each other.

The calculations show that as the module is reduced, even ignoring all the other problems typical of small and micro-drives, the different methods give increasingly different results. This is no longer the case in the module range above 1 mm.

However, it is not possible to determine by calculation which method gives the correct result, experiments are needed. Therefore, the next step of our research is to carry out a series of experiments to show which calculation method is the most appropriate for small plastic gears and whether the application of known methods is even practical.

Literature

[1] Schumann A.: Tribologische Untersuchungen von thermoplastischen Kunststoffen mit Modellprüfständen (https://www.triboplast.de/tribologische-untersuchungen-von-thermoplastischen-kunststoffen-mit-modellpruefstaenden/) (visited: 2021.12.26.)

[2] Antal M.: Műanyag fogaskerekek, Műanyagok gépészeti alkalmazása II. kötet, pp: 482 – 584. Gépipari Tudományos Egyesület Kiadványszerkesztési és Műszaki Iroda, Budapest, 1986.

[3] LICHARZ GMBH: Designing with engineering plastics. Licharz GmbH, Buchholz, Németország, 2016.

[4] VDI 2736 Thermoplastische Zahnräder, Blatt 2, BeuthVerlag, Berlin, 2014.

[5] JIS B 1759:2019 (JGMA/JSA) Estimation of tooth bending strength of cylindrical plastic gears, Japanese Standards Association, Tokyo, 2020.

[6] Hooke C. J., Mao K., Walton D., Breeds A. R., Kukureka S. N.: Measurement and Prediction of the Surface Temperature in Polymer Gears and Its Relationship to Gear Wear, J. Tribol. Jan 1993, 115(1): 119-124 (6 pages), 1993.

IOP Conf. Series: Materials Science and Engineering 1237 (2022) 012008 doi:10.1088/1757-899X/1237/1/012008

Redesign of a Volkswagen Crafter vehicle to a hybrid vehicle having e-motor and diesel engine

Sándor Bodzás[1], Zsolt Tiba[2], Piroska Ailer[3], Géza Husi[4]

[1]deputy head of department, associate professor
Department of Mechanical Engineering, Faculty of Engineering, University of Debrecen, Hungary

[2]collage professor, Department of Air- and Load Vehicle, Faculty of Engineering, University of Debrecen, Hungary

[3]vice rector, collage professor, Department of Air- and Load Vehicle, Faculty of Engineering, University of Debrecen, Hungary

[4]dean, head of department, associate professor, Department of Air- and Load Vehicle, Faculty of Engineering, University of Debrecen, Hungary

[1]bodzassandor@eng.unideb.hu, [2]tiba@eng.unideb.hu, [3]ailer.piroska@unideb.hu, [4]husigeza@eng.unideb.hu

Abstract
The aim of this study is to show the results of the vehicle research group redesigned a Volkswagen Crafter diesel type van to a multi hybrid vehicle.
The original chassis of the four-wheel drive van provides the base of the vehicle. The electric motor and the gearbox designed by the research group is placed in front of the rear suspension. This electric drive unit is coupled to the rear differential via the Haldex clutch. During the electric drive the transaxle of the van is in neutral gear and the diesel engine is not operating. Releasing the Haldex clutch the electric drive is detached from the vehicle and the diesel engine is activated to drive the front-wheels. Our purpose is to test the electric drive of the vehicle by conducting measurements on the electric motor and the battery, this way making new research directions.

1. Introduction

The Volkswagen Crafter (Figure 1), the first appearance is in 2006, has 3 to 5 ton mess in the largest case. The German car manufacturer, Volkswagen Commercial Vehicles, manufactured this brand. Originally, this type changed the Volkswagen Transporter LT nameplate first launched in 1975. The first issue Crafter was made in the Mercedes-Benz Ludwigsfelde and Düsseldorf plants. This is the same German factory where the Mercedes-Benz Sprinter is manufactured and assembled. A French car designer Laurent Boulay is responsible for the frontal design of the Crafter, which takes advice from the Volkswagen Constellation [1].

All the internal combustion engines were based upon Volkswagen Group's reengineered 2.5 litre R5 TDI between 2006 and 2010. This engine is an inline-five cylinder (R5) Turbocharged Direct Injection (TDI) diesel engine. It displacement is 2,459 cm^3 (150.1 cu in), and it has the latest common rail fuel system, with piezoelectric actuated injectors for the cylinder-direct fuel injection. It also use a diesel particulate filter (DPF), permitting all engine variants to comply with Euro IV European emission standards [1].

The 2.5 TDI CR engine was redesigned in the 2010th version because the previous problems of turbo failure had to be corrected. An update on the engine was also released, under the guise of "Blue TDI", which used AdBlue - or diesel exhaust fluid (DEF) - in combination with a DPF to gain the striker Euro V EEV European emission standards [1].

The Crafter has three wheelbase options: 3,250 mm, 3,665 mm and 4,325 mm. It is contained with front airbags as standard, and side and curtain airbags as options, along with anti-lock braking system (ABS), load adapting electronic stability programme (ESP), anti-slip regulation (ASR), and electronic differential lock (EDL).

The electronic differential lock (EDL) used by Volkswagen is not a differential lock at all. Sensors monitor both roadwheel speeds go through a driven shaft, and if one is rotating materially faster than the other the EDL system instantaneous brakes it. This potently forwards the torque to the other driven wheel that is regarded for still have grip [1].

Figure 1. The establishment of the Volkswagen Crafter [2]

The aim of our research is to insert an electric motor (e-motor) to drive the rear axes. In this case the front-wheels can be driven by the diesel engine and the rear wheels by the e-motor (Figure 2).

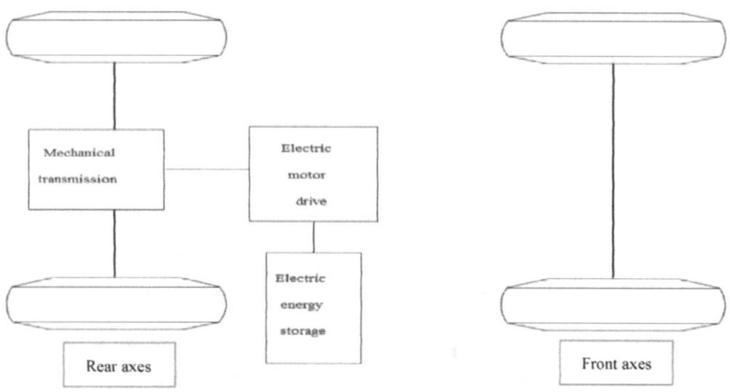

Figure 2. Integration of the e-motor

2. The assembly process and the usage of the vehicle

The vehicle can operates in two modes: diesel mode and electric mode. The required mode can be selected by a mode switch before starting the engine. During the ignition the mode switch is ineffectual.

2.1. Diesel mode

The vehicle is driven by the diesel engine. This engine follows the speed of the vehicle. It does not permit higher speed than 20 km/h.

2.2. Electric mode

The prerequisite of activating the electric mode:

- Switching on the ignition the e-drive makes a test. During this time the brake pedal has to be pressed. During the test if we do not press it or release it, the test will be interrupted and the e-drive can not be switched on. The restart can be possible based on out-in switch of the ignition switch again.
- The gearshift should be in neutral position.

2.3. Motion of the vehicle by e-mode

The forward mode is activated by the joystick moved to forward position. The middle position is the neutral. Switching back is the reverse gear. The condition of switching into forward or reverse gear is pressing the brake pedal. Switching into reverse gear is possible in case of a standing vehicle. The vehicle speed is limited for 20 km/h in e-mode.

2.4. Controlling of the e-vehicle

The Vehicle Control Unit (VCU) can get supply voltage from the ignition switch or from the charger port. The charging mode is activated if the supply voltage is received by the charger port. If it is not made, the VCU investigates the position of the mode switch. Depending of this position, diesel or electric mode will be activated.
During operating the VCU does the following inspections:

- the CAN communication between the components of the system,
- the function of the gas pedal,
- the function of transmission and number of revolution,
- the correct combination of the switching level,
- the voltage of the accumulator,
- the electric insulation by monitoring insulation resistance.

2.5. Design of a gear box for the e-mode

A new gear box was designed for a given electrical motor. The parameters of the engine are I_{max}=418/166 A, P_{max}=82.3/53.5 kW, U=320 V. The geometric parameters of the gears are on Table 1. The schematic figure and the CAD model can be seen on Figure 3 and 4.

Table 1. The initial parameters of the gears

Diameters, d [mm]	**Axial module, m [mm]**	**Number of teeth, z**	**Gear ratio, i**
d_1	2.25	26	1.54
d_2	2.25	40	
d_3	2.5	20	3.4
d_4	2.5	68	
d_5	3	18	2.33
d_6	3	42	
Σi=12.22			

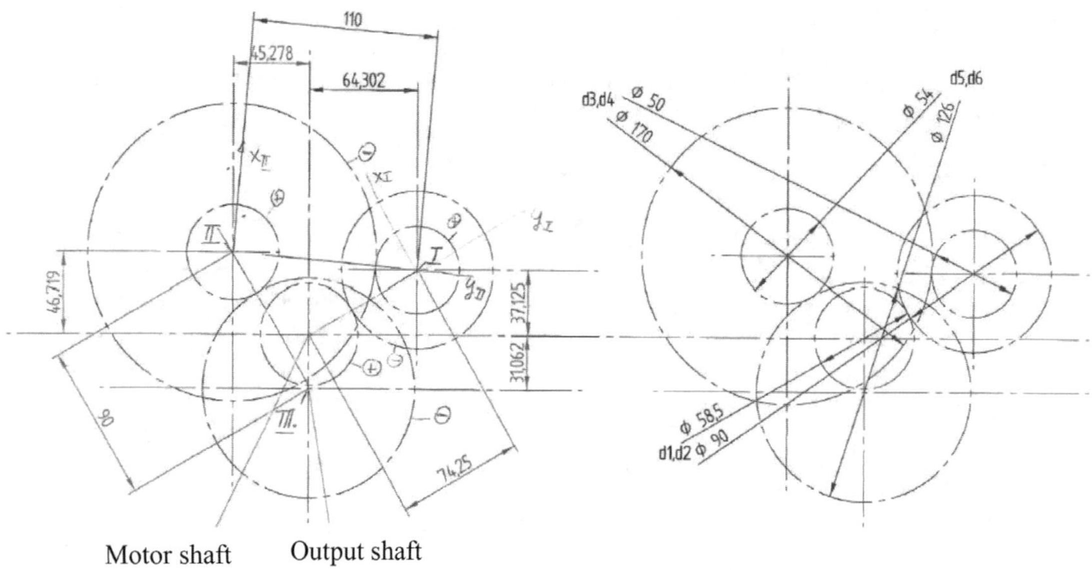

Figure 3. The arrangement of the gears into the gear box

Figure 4. The simulated models of the designed gearbox

Figure 5. The position of the gearbox at the bottom of the vehicle

IOP Conf. Series: Materials Science and Engineering 1237 (2022) 012008 doi:10.1088/1757-899X/1237/1/012008

After the geometric design the gearbox was manufactured and assembled into the vehicle (Figure 5). The build-up of the vehicle after the conversion can be seen on Figure 6. The electric control system can be seen above the rear axle on the vehicle platform.

Figure 6. The build-up of the vehicle after the conversion

3. Modelling and simulation using Permanent Magnet Synchronous Motor

The usage of the Internal Combustion Engine (ICE) in conventional cars is one of the highest reason of global warming and air pollution. The emission of toxic gases is also injurious to living body. Electric drive was developed in modern electric cars to change the ICE. A rear-wheel driven electric powertrain based on a Permanent Magnet Synchronous Motor was established. It can change its front-wheel driven diesel engine in the environment of the city using of low average speeds. A Nissan Leaf battery with a nominal voltage of 360 V and a 24 kWh power was simulated due to the application as the energy source of the electric drivetrain [3].

The New European Driving Cycle was utilized in this work to estimate the electric drive. Another input parameter is the speed enhancement because of the analysis of the vehicle on different road conditions. A Proportional Integral Controller was controlled the speed of the vehicle and synchronous motor. Different driving cycles were applied to test the vehicle [3]. Some simulation results can be seen on Figure 7.

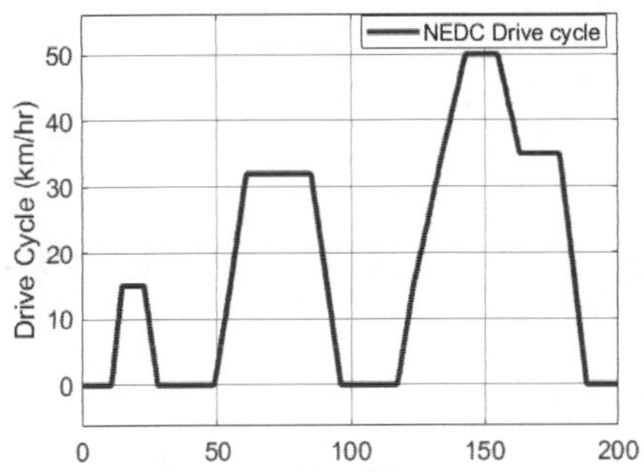

a) The speed gained by the vehicle

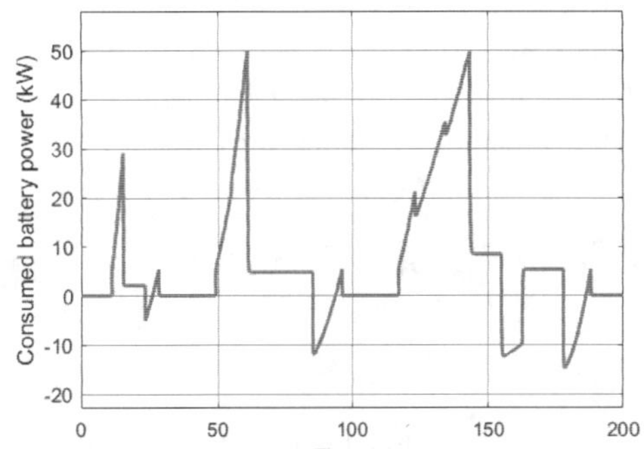

b) The power consumerism of the battery

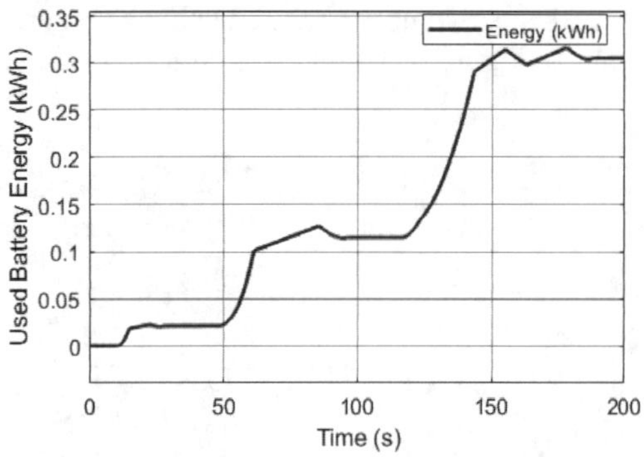

c) The energy consumerism of the battery

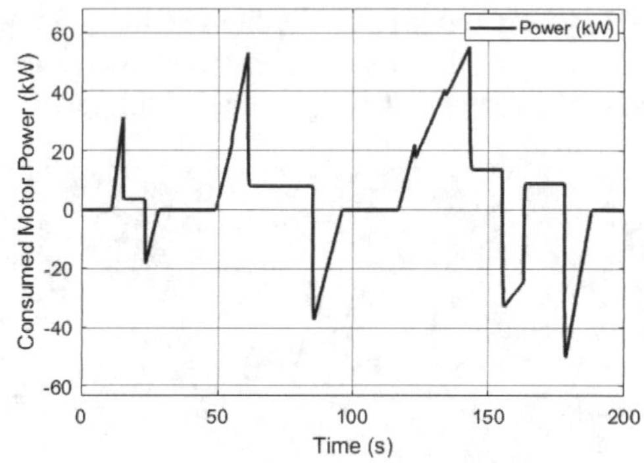

d) The power consumerism of the motor

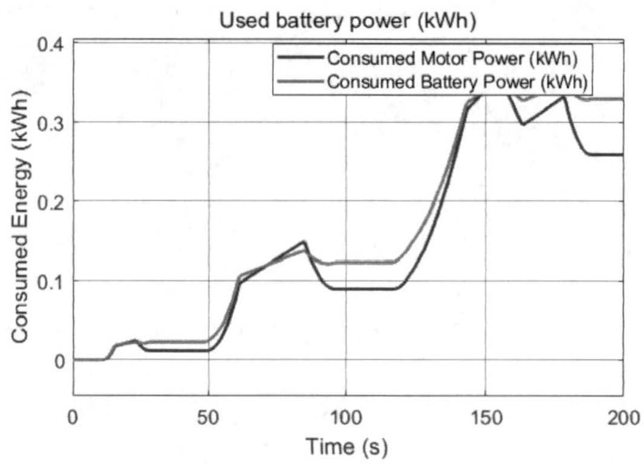

e) The energy consumerism of the motor and battery

Figure 7. The simulation results of the motor and the battery [3]

4. Conclusion

A Volkswagen Crafter diesel type vehicle was redesigned to a multi hybrid vehicle that can be operated by diesel or electric motor. The front-wheel drive is driven by the diesel engine. Switching into e-mode the rear wheel drive is activated and driven by the e-motor.

A new gearbox was designed to make a connection between the e-motor and the rear wheel drive.

A Proportional Integral Controller is used to get relation in connection with the speed of the vehicle and synchronous motor. We conducted simulations and modelling analysis to analyse the function of the e-motor and the battery under certain time.

Our purpose is to test the electric drive of the vehicle by conducting measurements on the electric motor and the battery, this way making new research directions.

Acknowledgement

Project no. **TKP2020-NKA-04** has been implemented with the support provided from the **National Research, Development and Innovation Fund of Hungary,** financed under the **2020-4.1.1-TKP2020** funding scheme.

References

[1] https://en.wikipedia.org/wiki/Volkswagen_Crafter

[2] https://www.truck1.eu/vans/flatbed-vans/volkswagen-crafter-2-0-tdi-143p-grijs-metalic-ac-4-a2577948.html

[3] **Babangida, A., Szemes, P. T.:** *Electric vehicle modelling and simulation of a light commercial vehicle using PMSM propulsion*, Hungarian Journal of Industry and Chemistry, Vol. 49(1) pp. 37–46 (2021), DOI: 10.33927/hjic-2021-06

Conversion of a Lada 2106 car into a rally racing car

Ferenc Zombori, Sándor Forrai, István Péter Szabó

University of Szeged Faculty of Engineering
Moszkvai körút 9. HU6729 Szeged, Hungary

pszi@mk.u-szeged.hu

Abstract. The subject of this publication is a rally car designed as a diploma thesis and later produced. G-Sport Ltd. designs and manufactures racing car bodies, chassis components and all other motorsport related components for both the domestic and international markets. The aim was to design the Lada 2106 race car to accept a Ford Mk2 suspension, a Toyota 4A-GE engine and other parts that would enable it to compete in official FIA rally events. The work was done using a 3D scanner. The first phase involved creating 3D models of the body, engine, gearbox and other components, while checking their dimensional accuracy. Then, the custom parts of the car, including the roll cage, had to be designed to meet FIA specifications. This article describes the steps of the 3D scanning method, how the scans are processed, oriented, assembled correctly, and how the deformations of the body frame were detected. In addition, the typical problems encountered during 3D scanning of large objects and their generally applicable solutions are shown. The components of the drivetrain and components of own custom design are also described.

1. History

G-Sport Ltd. designs and manufactures racing car bodies, chassis parts and all other parts related to motorsports for both domestic and international markets. The design and manufacture of a prototype car in 3D was under preparation. The plan was to redesign a 2101 Lada body to adopt a Ford Mk2 chassis, a Toyota 4A-GE engine and other parts, which would enable it to compete in an official FIA rally race.

3D scanning is used in the automotive industry in many areas, especially for modifying vehicles, e.g., in the manufacture of sightseeing buses or racing cars [1]. Improving aerodynamics can be one of the goals when making racing cars. 3D scanning is an important tool for the IndyCar field to create a digital model of previous racing cars [2]. The method is also used successfully in Formula 3 racing cars. An Artec Leo scanner was used to scan the DallaraF399/01 race car [3]. These scanned models can be used to verify the original CAD models; an important starting point for further development.

Scanning a car can run into several difficulties. This can be caused, for example, by shiny, smooth surfaces. Glass surfaces should be covered or temporarily painted, glossy surfaces should be patterned [4], or other methods detailed in this publication should be used to assist the process. Among Artec's scanners are those that are specifically designed to scan large objects, yet are accurate enough for the job [5] [6]. They are very effective tools when rebuilding cars, significantly speeding up the design process. This is true not only for body modifications but also for engine, driveline, or chassis modifications [7] [8].

2. Requirements for the racing car

The set of prototype cars is called the group P. This may include cars that combine the features of numerous cars. In Hungary, the MNASZ (Hungarian National Motorsport Association) has issued a regulation for the existing FIA (International Automobile Federation) regulations, entitled "Specific National Rules for Group P Vehicles". In this document, it defines 3 subcategories for prototype cars: P - rally, PH - Rally Hobby-car, PT - Terrain Rally [9]. Article 253 of the FIA Appendix is a set of rules issued by the FIA that contains rules for safety equipment for cars in groups N, A, R-GT. The appendix deals in detail with the roll cage and its design. In accordance with the rules, roll cages could receive ASN (National Automobile Association) homologation and must be marked with an identification plate, which must include the manufacturer's name and a unique serial number.

The roll cage is a tubular structure inside the passenger compartment, that is installed close to the chassis and is intended to reduce the deformation of the body in the event of an impact.

3D scanning is an opportunity for reverse modeling of 3D elements and components, in which information about more complex components that are difficult to model, such as an engine block, steering knuckle, etc. is obtained.

During the work, several such recordings had to be made about the different parts, which can be basically divided into two groups. The first group includes the chassis itself, which does not have to move, its position can be considered fixed. The second group includes parts that can be considered more general, such as the steering gear.

3. 3D scanning of the 2101 chassis

An Artec EVA 3D scanner was used for the work [10]. The design of the car required the entire chassis (Figure 1) in order to design the fitting parts, for example: the roll bar, the shock towers, and to visualize the relative position of the individual parts, such as the engine position in chassis.

Figure 1. A 2101 chassis on a rotisserie

The chassis was sandblasted so that the surfaces were non-reflective, which was essential since bright surfaces could not be scanned.

IOP Conf. Series: Materials Science and Engineering 1237 (2022) 012009 doi:10.1088/1757-899X/1237/1/012009

An internal shot was taken that included the chassis viewed from the passenger compartment, to which the engine compartment also belongs viewed from above, as well as the luggage compartment also viewed from above.
This interior room was divided into 1meter x 1meter sections. Such large areas were scanned simultaneously so that there was always an overlap between the shots.
The device is able to scan much larger areas in one piece, however, in this case the accuracy of the images drastically reduces. This 1meter x 1meter raster is an experiential value with an accuracy of ± 0.2 mm at this size.
After the recordings have been made, the magnitude of the error number gives a preliminary indication of the quality of each scan.
The roof, pillars A and B were more difficult to scan. The roof was only scanned from the inside, as it only encountered matching parts from the inner side. EVA loses its orientation in space on large flat surfaces, as a result, the recording becomes unusable. For this reason, magnets were placed on the flat plate of the roof, which served as reference points for EVA. The roof was a big challenge, even with the magnets shown in Figure 10; the shots weren't always good, so a piece of clothing was placed on both sides of the scanned areas (Figure 2). Its irregular geometry ensured that EVA did not lose its orientation.

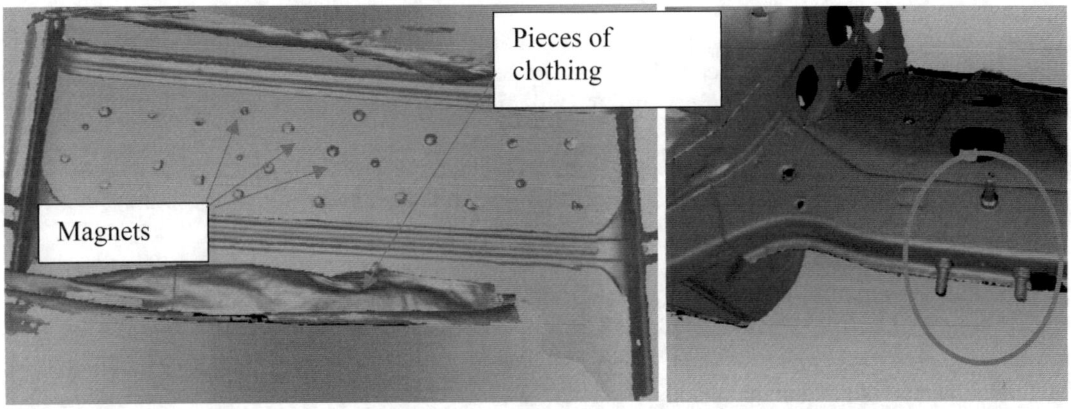

Figure 2. Rough view of the center of the roof with scanning aids and screws, defining the axis line of the holes

Scanning pillars A and B was a problem, as the scanned parts were extremely narrow, thus EVA lost its orientation. To eliminate this, the chassis was turned to its side and extra objects were placed next to pillars A and B, therefore increasing the size of the scanned surfaces.
Before scanning the lower part of the chassis, it had to be considered which points would be the ones that would be used in the future. Such points are, for example, the front cradle points and the rear trailing arm connection points. Screws were tightened to these points. That is how they were visible on the shots. Images of their axis lines were taken later (Figure 2). Scanning was performed in this case too as it was described above.

4. Processing and compilation of the recordings of the 2101 chassis

The scan was followed by the processing of the recordings, which was done in Artec Studio. 3. Different algorithms had to be run on each recording to get the mesh files with the final size and resolution. In the first step, algorithms had to be run, which fine-tuned the size and position of the recordings (Rought serial registration, Fine registration, Global registration).
The software is able to correct 5-7 mm slips, but unable to correct larger ones, such shots are useless (Figure 3).

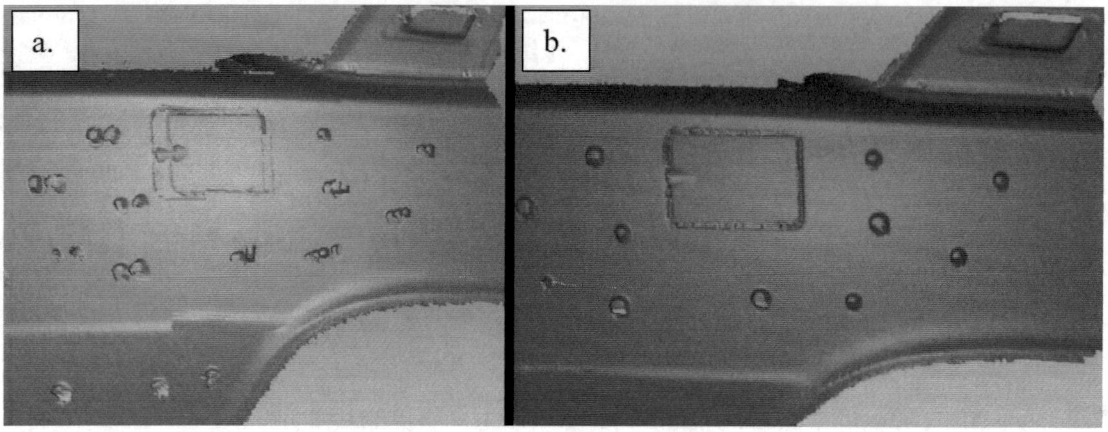

Figure 3. "Shifted"/unusable (a) and good (b) recordings of the right rear fender

In the second step, the algorithm that cleans the edges and contours of the recordings and eliminates the disturbances on the surface of the recordings (Outlier removal) was run.
The last step was to run the "Fusion" algorithm. This finalizes the recording, which "fuses" the images into one layer.
The first algorithm, based on the orientation data, arranges these images in the right place, the second algorithm cleans the surfaces, removes disturbances, and the third one averages these images into one "image".

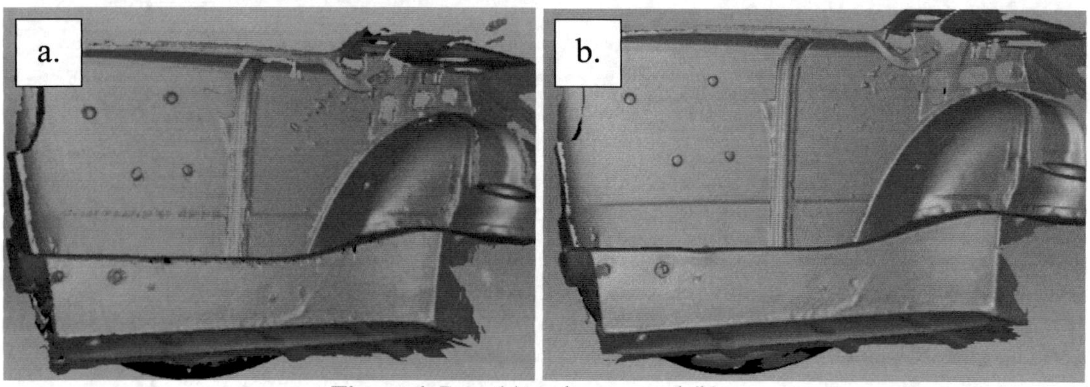

Figure 4. Raw (a) and processed (b) scan

After cleaning, 1m x 1m processed shots were created. The next step was to make a final scan out of these shots.
2 recordings can be composed at the same time. The origin of the first recording, to which other recordings are joined, will be the reference origin. While joining the recordings, reference points are being looked for on the images (Fig. 5).
Having joined all the recordings together, the "Registration" and "Fusion" algorithms had to be run again, which re-arranged and averaged the individual recordings. The last step was to examine the dimensional accuracy of the whole recording in relation to reality.

Figure 5. Joining 2 recordings of the right rear drum plates with reference points

At this point, point pairs were also being looked for that were well measurable from each other. Special attention was paid to critical locations during the work, such as the distance between the front spars, and the distance between the inner sill edge and the roof (Figure 6).

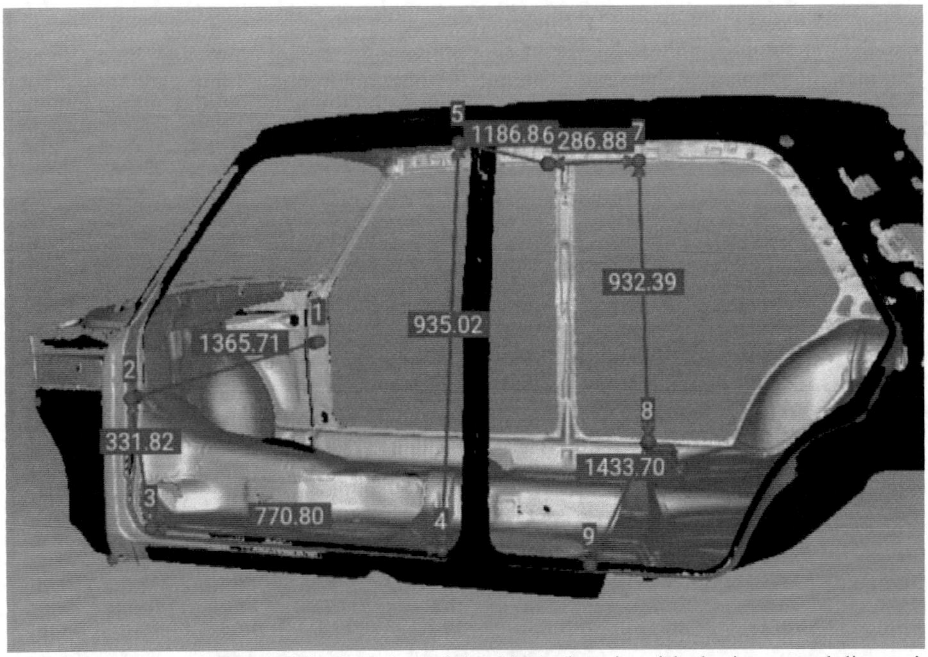

Figure 6. The completed image of the interior of the 2101 chassis with the inspected dimensions on it

The mesh recording of the 2101 chassis has been completed, which consists of 2 main parts as described above. They are "Internal Recording" and "External Recording." In order to perform substantive technical work with any mesh, orientation needs to be executed on them.

In each case, the orientation was performed in SolidWorks, in a module called SolidWorks for Geomagic. If possible, adjustments must be done to those points on the chassis, whose position is clear and well recognizable. Examples of such points on the chassis are the front cradle points, rear cradle points, trailing arm connection points, etc.

By arrangement, the origin had to be placed in the middle of the back points of the first cradle. During the scanning, screws were tightened into the first cradle points. Screws were put into the trailing arm

connection points as well. With the module called Geomagic, a plane was recorded on the bearing surface of the points of the first cradle.
Subsequently, the axis lines of the screws, which were screwed into these points, were recorded too in such a way that in the recording it was specified, that these axis lines had to be perpendicular to the previously recorded bearing plane. The same process was done at the trailing arm connection points on the rear of the chassis. As a result, 4 points were obtained on the chassis. In a 3D sketch, the first two points and the back two points were connected. The centers of these lines defined the center plane of the car.
Before scanning the chassis, it was found out that the right front of the car could have been damaged in the past because there were traces of repair work there. The spars were examined in the mesh, and it became clear that they were significantly twisted.

5. Engine scan

The Artec EVA 3D scanner was also used to scan the engine. The engine was scanned together with the gearbox housing. With the help of colleagues, the gearbox housing and the structure of the clutch was removed from the engine. Pictures about it were taken in this form. This was necessary because it made the flywheel visible; the significance of which will be explained during the orientation.
The engine will be adjusted into the chassis model along with the axis line of the crankshaft. One of the most important aspects was to be able to capture this axis line as accurately as possible. This is why the flywheel was needed, because the axis line of the crankshaft could be defined by using its cylindrical surface (Figure 7). From the polygons marked in red, the system calculates the axis line which was specified to be perpendicular to Plane1 recorded in the rear plane of the flywheel.

Figure 7. Recording of the axis line of the Toyota engine main shaft based on mesh

The placement of the engine and transmission unit was emphasized in terms of positioning the largest mass in the car, which was very important for the final center of gravity. It had to be placed as low and as far back as possible. In a suspended position, it was examined what the lowest position of the car would be to the ground and then its height was adjusted by leaving a safety distance. The other essential matter was to keep the engine, transmission, and its environment repairable with enough space to access individual parts even in competitive conditions.

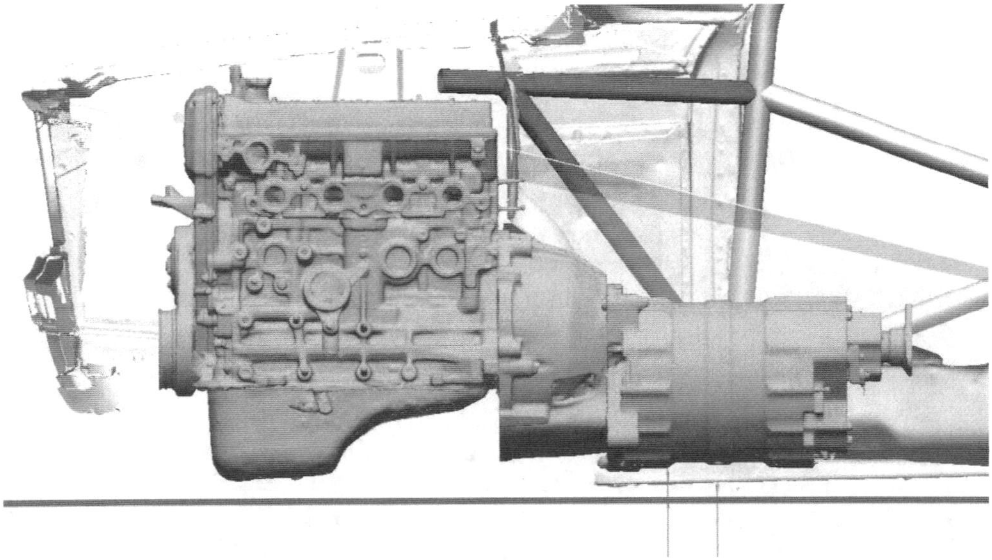

Figure 8. Positioning of the engine-gearbox unit in the chassis

6. Rear axle orientation

The orientation of the rear axle was relatively simple due to the nice regular surfaces. The orientation started by recording the axis line of the output shafts by using Geomagic, so that essentially axis X was defined (Figure 9).

Subsequently, the bearing plane of the hubs, the bearing plane of the output shafts and its axis line were recorded.

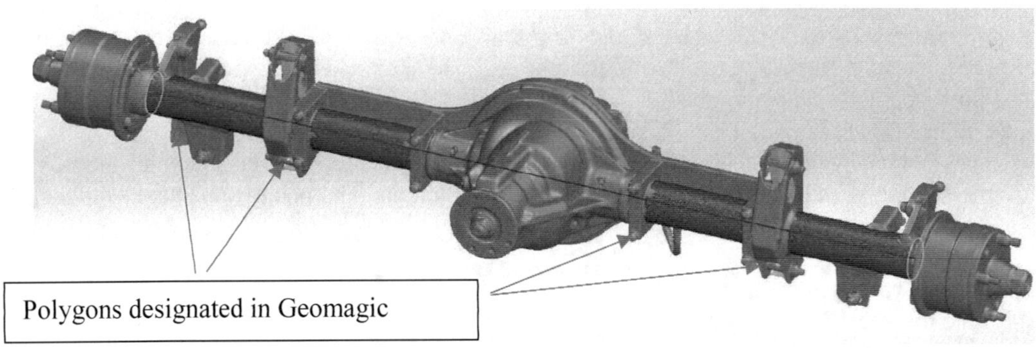

Figure 9. Recording of the axis line of the Ford rear axle output shafts based on mesh

7. Floor Tunnel

The floor tunnel is also subject to the rules laid down by the FIA (International Automobile Federation). The entire tunnel is made of high-tensile Docol 800DP plane.

IOP Conf. Series: Materials Science and Engineering 1237 (2022) 012009 doi:10.1088/1757-899X/1237/1/012009

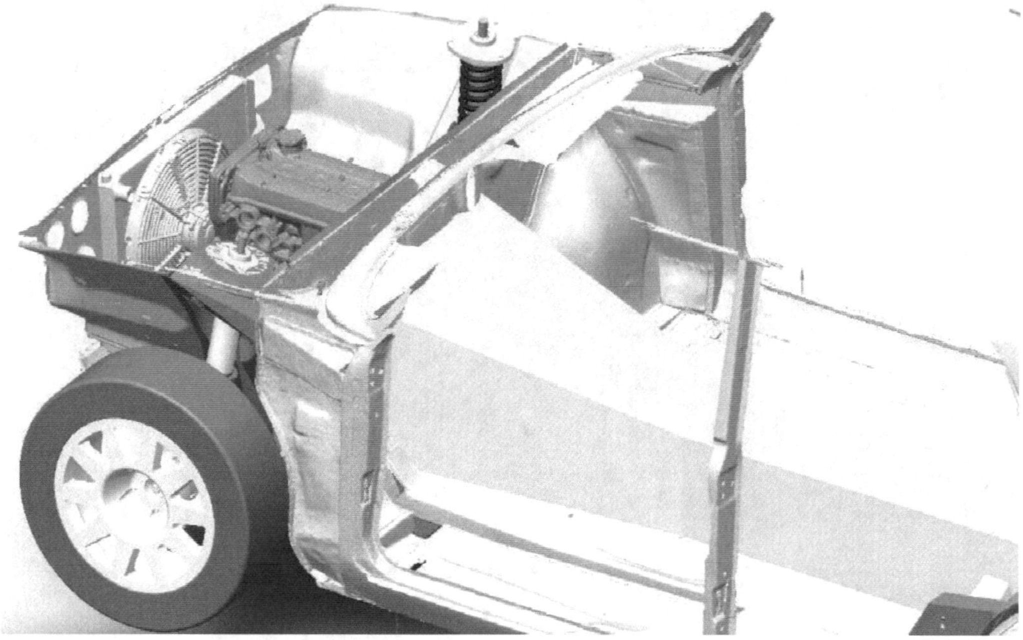

Figure 10. Parts of the floor tunnel

While drawing the floor tunnel, the plate was overhung everywhere on the chassis. The part was cut to the correct size with the connecting surfaces of the scanned chassis. The 3D scanning thus allowed the laser-cut and the bent plane metal parts to fit precisely, without further alignment.

The location of the seat had to be considered, while designing the floor tunnel. As far as the seating position was concerned, it was relevant that the passengers of the car sit as far back as possible in order for the center of gravity to be optimal. Another important aspect was that they could look out of the car and be able to position themselves comfortably. The seat and seat holder are also regulated by the FIA. The seat position had to be determined in such a way that the seat belts approach the seat at the angles specified in the rule [11].
The seat position was followed by the steering pillar and pedal position. To maintain the center of gravity, the steering wheel and pedals are adjustable, not the seat. The pedal was attached to a sliding rail.

8. Design of the roll cage

The roll cage in the car is a safety unit, therefore it is a highly regulated area by the FIA [3]. In relation to the weight and performance of the car, the size and tensile strength of each pipe are given. Among others, it is also given how much each node can be displaced due forces on the roll bar.
By mutual agreement, a roll cage with a rear main hoop was designed into the car. In accordance with the regulations, the recording of the specified load forces is shown in Figure 11.

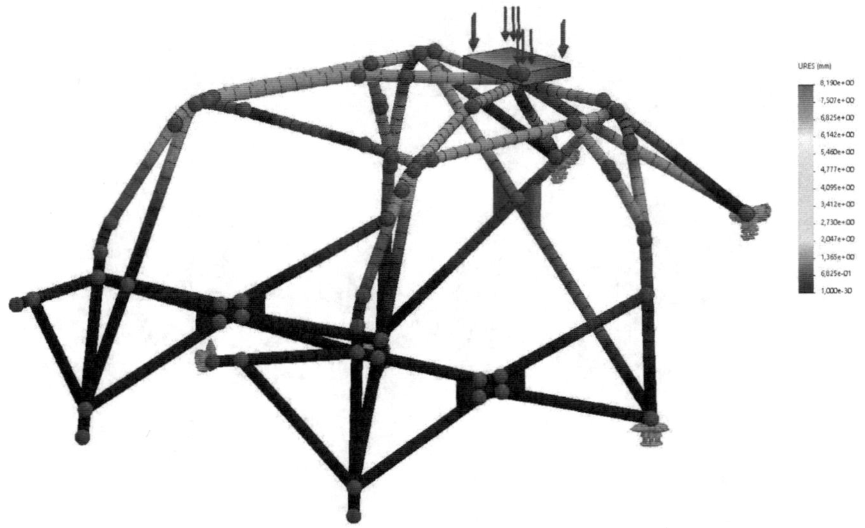

Figure 11. Analysis of the vertical static load of the roll cage at the main hoop

Under vertical load, the displacement of the node of the main hoop under a force of 78.75 kN was 8.1 mm. The maximum allowable displacement here, can only be up to 50 mm. (FIA, 2017)
In the second analysis, the largest displacement according to the study was 2.5 mm under a force of 36.75 kN (Figure 12). Here, in this case, the maximum allowable displacement is 100 mm. (FIA, 2017)

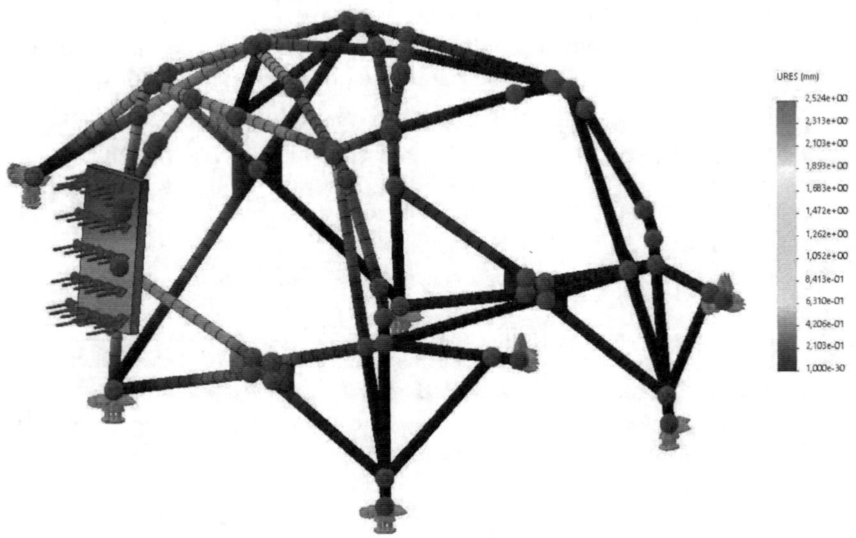

Figure 12. Analysis of the horizontal static load of the roll cage at the main hoop

Test 3 was performed at the lateral half-roll bar. At this time, the magnitude of the load force and the dimensions of the pressure prism did not change compared to the previous test (Figure 13).

IOP Conf. Series: Materials Science and Engineering 1237 (2022) 012009 doi:10.1088/1757-899X/1237/1/012009

Figure 13. Static load test of the roll cage at the lateral half-roll bar

In this case, the maximum displacement was 2.49 mm under a force of 36.75kN. As in the previous test, the maximum displacement could have been up to 100 mm. The results of the three tests produced a relatively low value, which can be best explained by the small weight of the car and the magnitude of the load force.

The designed roll cage passed the load test, did not prove to be too rigid, and complied with the values specified in the rule. This roll bar was made of 25CrMo4. The weight of the completed roll bar was 40.5 kg. While designing the roll bar, the material quality, and sizes of it were different to the parameters given in the FIA regulation. The primary reason for this was to lose weight. In order to prove our design, a version of the roll cage was made using the mechanical properties, bar sizes and node displacements specified in the rules. It is important that a minimum tensile strength of $350\,\frac{N}{mm^2}$ is given in Appendix J. An ST35 (ANSI 1020) bar was used for this roll cage version with a tensile strength of $420\,\frac{N}{mm^2}$.

9. Summary

Plans for a race car based on the Lada body have been completed. We used modern methods for the design. 3D scanning significantly facilitated the production of some parts, so no subsequent modifications were required. According to our plans, the build of the racing car is in progress. The custom body kit for the race car is currently being developed.

Figure 14. Interior of the racing car under construction

The results can be used when designing additional racing cars. A roll cage with sufficient strength but reduced weight is an advantage in competitions. The reverse engineering method, which uses a scanned body model, greatly simplifies the design and manufacturing process. It can be used to accurately detect previous body defects and the deformations they cause. Scanning the body in 3D is an unusual task, given that the large part contains relatively small, plate-thick dimensions, so the accuracy of the final result is determined by the accuracy of assembling each scanned part. The methods used for this purpose are also presented in the publication.

References

[1] Lulić Z., Tomić R., Ilinčić P., Šagi G., Mahalec I. (2013) Application of Reverse Engineering Techniques in Vehicle Modifications. In: Stjepandić J., Rock G., Bil C. (eds) Concurrent Engineering Approaches for Sustainable Product Development in a Multi-Disciplinary Environment. Springer, London. https://doi.org/10.1007/978-1-4471-4426-7_78
[2] https://www.racecar-engineering.com/articles/indycar-3d-scanning/
[3] https://metrology.news/3d-scanning-data-enhances-race-car-aerodynamics/
[4] https://3dscanningservices.net/project/3d-scanning-a-race-car-for-computational-fluid-dynamics/
[5] https://www.artec3d.com/cases/tekk-consulting-auto-chassis-design
[6] https://www.laserdesign.com/high-precision-3d-scanning-using-artec-leo-for-custom-race-car-design
[7] Blagojevic, Milan - Bojovic, Milan - Milojević, Saša - Marković, Petar - Lazarević, Dragan: MODIFICATION OF RACING CAR CYLINDER HEAD USING 3D DIGITIZATION AND REVERSE ENGINEERING
October 2020
Conference: 8th International Congress Motor Vehicles & Motors 2020At: University of Kragujevac Faculty of Engineering
[8] Voicu, A. C., Gheorghe, I. G., Badita, L. L., & Cirstoiu, A. (2013). 3D Measuring of Complex Automotive Parts by Multiple Laser Scanning. In Applied Mechanics and Materials (Vol. 371, pp. 519–523). Trans Tech Publications, Ltd. https://doi.org/10.4028/www.scientific.net/amm.371.519
[9] https://mnasz.hu/download.php?doc=1536
[10] http://docs.artecgroup.com/as/13/en/_downloads/9817bfb9a490872fd8de27b7575c08d2/Manual-13-EN.pdf
[11] https://mnasz.hu/download.php?doc=5108

IOP Conf. Series: Materials Science and Engineering 1237 (2022) 012010 doi:10.1088/1757-899X/1237/1/012010

Eco-marathon racing car casing and frame structure design

Emese Kozák, Ádám Gégény, Ferenc Zombori, Sándor Forrai, István Péter Szabó

University of Szeged Faculty of Engineering
Moszkvai körút 9. HU6729 Szeged, Hungary

pszi@mk.u-szeged.hu

Abstract. In addition to the Pneumobil competition, our team will also participate in the Shell Eco-marathon with an electric-powered prototype category car. Based on our results so far, we are in the midfield of European competitions. After several modifications to our current car, we will be building a new car again in 2022. The goal of building eco-marathon racing cars is to achieve the lowest possible energy consumption, and accordingly, great emphasis should be placed on reducing weight and drag coefficient.

Our previous cars used a welded aluminum frame and fiberglass. One of the best ways of weight reduction would be a self-supporting, carbon-fiber composite frame structure; however, the team does not currently have the technology to produce this. The frame structure of the new car consists of carbon fiber composite tubes, the nodes are 3D printed, fiber-reinforced composite elements. The streamlined casing of the vehicle consists of vacuum formed PETG elements. By using this structure, the total weight can be radically reduced at minimal cost. In our publication, we describe the main steps in the design of the casing and frame structure, finite element analysis of air resistance and deformations, and the drive chain.

1. History of the Shell Eco-marathon competition

The Shell Eco-marathon is a global academic program, focused on energy optimization. It is one of the world's leading student engineering competitions. Over the past 35 years, the program has consistently brought to life Shell's mission of powering progress, by providing more and cleaner energy solutions. The global academic program brings together Science, Technology, Engineering and Maths (STEM) students from across the globe to design, build and operate some of the world's most energy-efficient vehicles. All in the name of collaboration and innovation, as students' bright ideas help to shape a lower carbon future for all. The competition dates back to 1939, when Bob Greenshields, Shell's Research Director, made a friendly wager with his colleagues to see who could get the best fuel economy from road cars adapted to maximize energy efficiency. Bob achieved 49 MPG, setting a record, and creating a legacy that has lasted the distance.

The Shell Eco-marathon officially launched in 1985, when 25 teams gathered in France, to pit their wooden vehicles against one another, in a format that laid the foundations for the competition we now know and love.

The track competition has come a long way since then and today's teams can choose from an array of energy options, from battery electric to hydrogen fuel cell, using different materials, from advanced carbon fiber to bamboo. In 2020 and 2021 with the cancellation of live events, the Shell Eco-marathon made the transition to a virtual program, resulting in great achievements, despite numerous odds. But one thing that hasn't changed is Shell's commitment to provide STEM students with a platform to collaborate and learn, to help shape a lower carbon world.

To date, more than 100,000 students from hundreds of universities in over 60 countries have participated and the legacy of the Shell Eco-marathon is clear to see. At these competitions one can meet the teams from previous and read their inspirational stories, with real-world impact on the STEM global community [1].

The general aim of the Shell Eco-marathon is to inspire teams of university students to build energy-efficient vehicles in various specified categories. Teams can design their vehicle in two categories: according to the specifications of a prototype or an urban concept. Within the two categories, vehicles with an electric or hydrogen-powered internal combustion engine can be built.

STECO – the team of the University of Szeged Faculty of Engineering enters the electrical category and has also taken part in the same category in previous years as well [2] [3] [4]. Although questions arise [5] [6], the spread of electric drive is becoming more and more intense today. An important advantage is the simplicity and low weight of an electric drive. The developments concern the control electronics of the electric motor.

The target is to build a vehicle that can run the longest distance possible by using the energy equivalent of 1 kilowatt hour.

2. Results of the team Szeged so far

The team of the University of Szeged Faculty of Engineering has been participating in the Eco-Marathon competitions in the electric-powered prototype vehicle category since 2015. Based on the results, the team is in the European mid-range. The best result of the team so far was 242 km/kWh at the 2017 Le Mans gala race, where this team earned the best Hungarian achievement in the competition.

Figure 1. STECO team, 2015

IOP Conf. Series: Materials Science and Engineering 1237 (2022) 012010 doi:10.1088/1757-899X/1237/1/012010

Figure 2. STECO team at the 2016 London race

Figure 3. London, 2019

The membership of the team is almost the same as that of the Airrari Szeged pneumobile team. At the Techtogether competition, which is organized twice a year, the team can claim numerous successful achievements, special awards, and category victories under the name of Airrari-STECO. The team's best result was the absolute first place of the 2018 GTE Techtogether competition.

Figure 4. Techtogether, 2021

3. Presentation of the racing car used from 2017 to the present

3.1. Frame structure
The frame of the car is a welded aluminum structure. The advantage of this was the simple manufacturing technology. Compared to the previous construction, it was possible to reduce the weight of the 2017 frame, while making the structure more rigid. The frame of the next car is designed to be much lighter with similar strength.
During the race, the car reaches an average speed of 25 km/h, rolling as smoothly as possible. The surface of the racetrack is smooth, good quality asphalt. For these reasons, a rigid suspension system is used.

3.2. Engine, drivetrain
The drivetrain consists of the following elements: electric motor - sprocket - chain - sprocket – a pair of wheels - freewheel. The ready-made, purchased elements of the drivetrain were selected in such a way that they have high-quality bearings, precise fitting, and careful design.
In 2017, the electric motor was a Maxon type with a nominal voltage of 48 V, a power of 250 W and an efficiency of 89% (at an operating speed of 3100 rpm).
The engine drives the rear wheel through a 180/22 chain drive with a transfer of 180/22. When choosing the type of chain, it was taken into account which chain the other teams use in the competition and which chain has sufficient tensile strength and the lowest possible weight. That is how the 04B chain, which is fortunately available in Hungary, was chosen.

3.3. Body
The body is a composite structure: polystyrene resin reinforced with fiberglass fabric. The body is not a load-bearing element, its only function is to reduce air resistance. To manufacture the body, the positive shape was created by cutting to size and gluing the polystyrene planes after printing the appropriate cross-sections.

Figure 5. Manufacturing the positive shape of the body of the 2017 racing car

The surface was coated with gypsum and after sanding and applying the mold release layer, the body was manufactured. In order to reduce weight, the body of the new car is manufactured with a different technology.

4. Rules for the body of the new vehicle
The designed vehicle must be built in accordance with the rules of the Shell Eco-marathon competition in order to participate in the event.
The 2021 regulation specifies that: the height of the vehicle must be less than 1000 mm; the rail gauge of the vehicle must be at least 500 mm; the quotient of the height and the width of the vehicle must be less than 1. 25; the axle-base must be at least 1000 mm; the maximum width of the vehicle must not exceed 1300 mm; the maximum length of the vehicle must not exceed 3500 mm. There must be a distance of at least 100 mm between the pilot's leg and the nose of the vehicle [7].

The final shape of the car could be formed through several concepts (Figure 6). Significant and conspicuous changes were made compared to the current racing car.

In contrast to the previous 2017 vehicle, all the 3 wheels of the new concept are located under the body, and the body of the new vehicle is provided with significantly larger rounding, as a result of this, the frontal area of the vehicle is reduced.

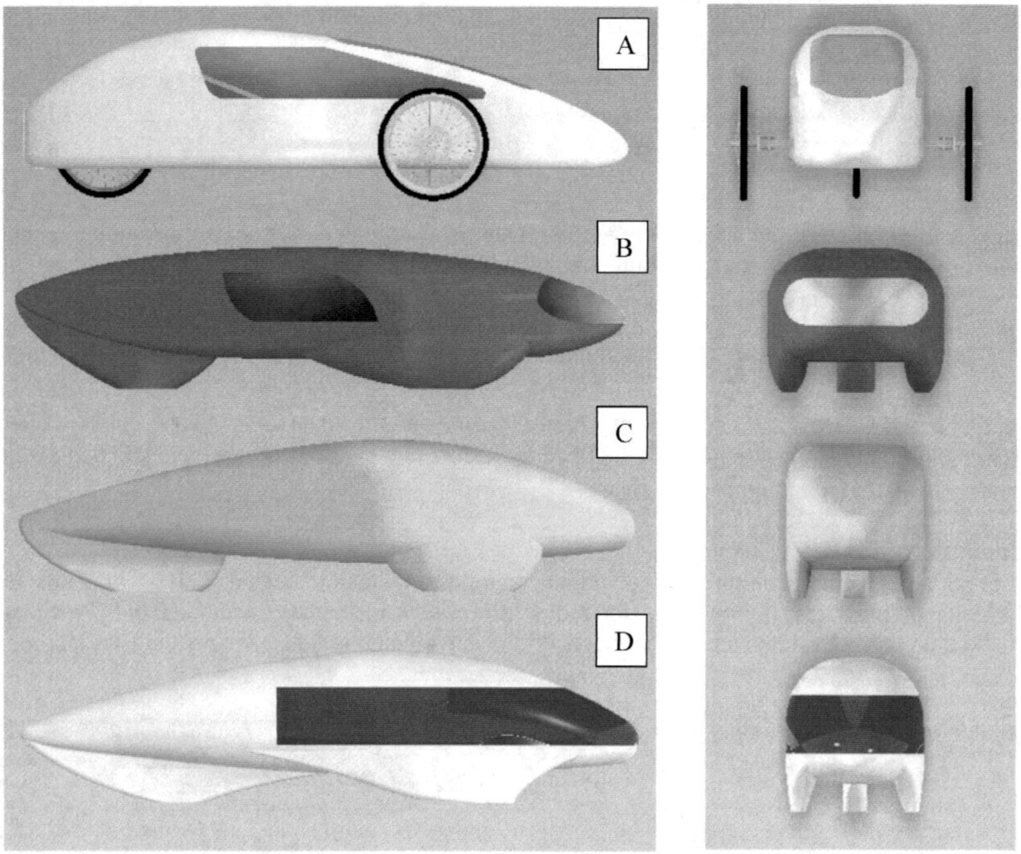

Figure 6. Comparison of the 2017 racing car and three new concepts: Version A, B, C, D.

In several steps, the version shown at the bottom of the figure was realized. The creation of this is described below.

5. 3D scanning of the pilot

Taking into account the requirements of the regulation, the vehicle must be designed to be the smallest and the most aerodynamic to embrace the pilot as much as possible. The best way to do this is to know the size of the pilot and build the chassis of the racing car by using this information. In order to achieve this, the design required a 3D 1:1 scale model of the pilot, around which the body could be designed.

A 3D scan was required to create the CAD model. The 3D scanning was performed with a scanner called Artec EVA Lite and its processing software called "Artec Studio" [8].

The pilot actually takes a "lying" position, in which they still know that they can drive, and then the scan takes place. After capturing the images, the program generates a 3D model, that can be converted into the CAD design programs.

Three shots were taken: two "full" shots and one with the head in high focus. Making these recordings barely took 5 minutes, with control accuracy. The number of errors associated with the particular recording in the processing software helped to verify the accuracy. This error number must be below 1 before processing the given point cloud. In this case, the recording is suitable for further work, if it is

IOP Conf. Series: Materials Science and Engineering 1237 (2022) 012010 doi:10.1088/1757-899X/1237/1/012010

above 1, then it must be recorded again. The next step is to join the recordings to the identified joint points (Figure 7).

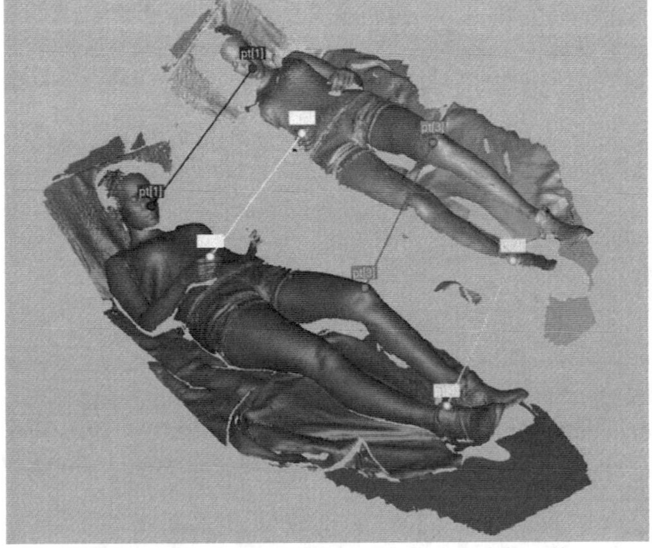

Figure 7. Joining the 3D recordings of the pilot

6. Making the 3D model of the vehicle's body

On the x-y plane a sketch is made that represents the top view of the body (Figure 8). The driver's inclusive sizes are made by guiding lines. The wheel position, the 500 mm gauge and the 1400 mm wheelbase are determined. The front wheel's position is planned for the driver's knee flexion. By comparison I made the lowest size wheelbase. In this case there are other things that matter, but the driver. In the engine compartment, there is the electric motor, the battery, the chain drive and the position of it.
The next step is a sketch on the y-z plane. It represents a side view of the vehicle. (Figure 9).

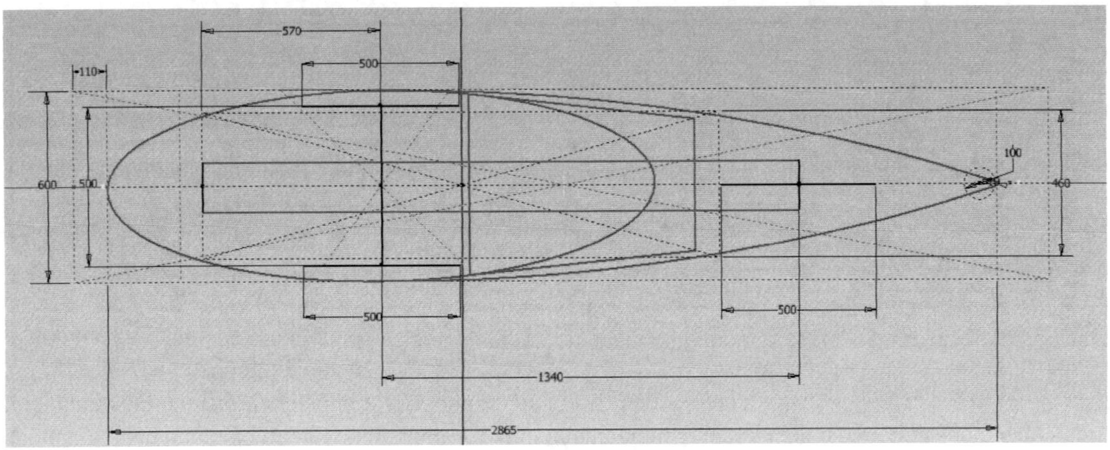

Figure 8. *Top view of the car body*

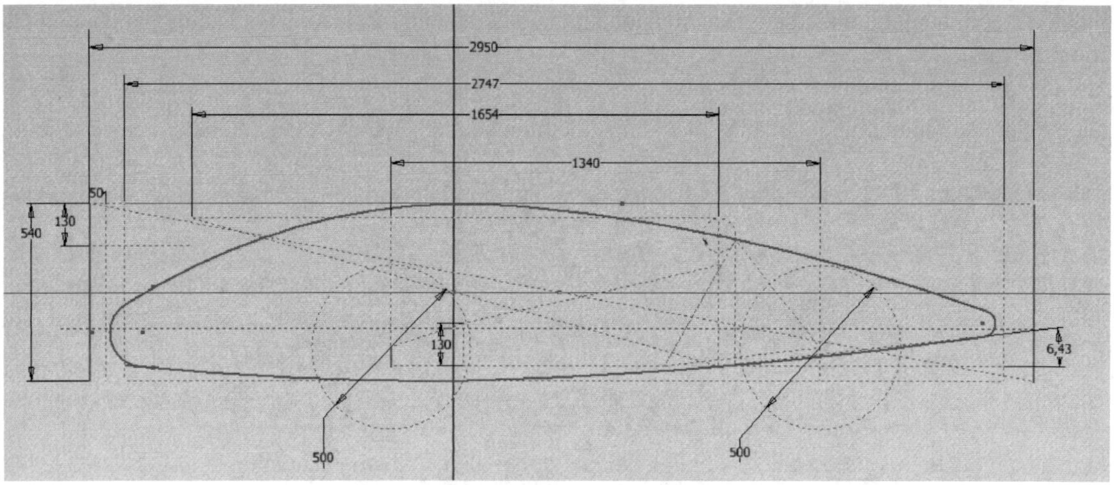

Figure 9. A *Side view of the car body*

When the sketch about the top and side view is finished on the x-y and z-y plane, the side view sketch is extruded by the „Extrude" command 800 mm long (the essence is to be longer than the full width, so it should be extruded further than 600 mm).

After all, by the Extrude command the top view is designated and extruded. The important thing about it is the extrusion is not a solid body extrusion, but also an engraving extrusion (on the designated sketch everything should be cut out).

After the cut out we get a „capsule" shape with sharp edges. These sharp edges should be rounded as much as possible. For the rounding the „Fillet" command's „Variable" option was used for rounding with radius changes. First the rounding was made in half of the body and mirrored for the y-z plane, so a symmetrical rounded body was created.

The body setup is finished so the cover of the wheels must be made. The wheel position is known. A work plane must set on the y-z plane. It should be moved 310 mm and a sketch should made about the front wheel's cover.

The sketch is extruded as a solid body 300 mm long. (Figure 10)

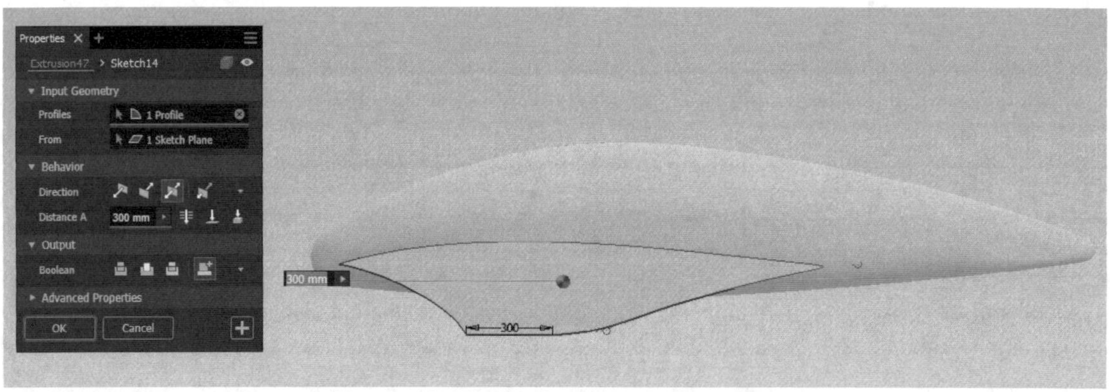

Figure 10. *Front wheel's cover extrusion*

For the perfect shape for the wheel cover, the „Loft" command was used and some thickening and tightening was made. For this command three planes were set at different heights and a sketch was made about the cross-section of the wheelbase. The Loft command's „engraving" option is used for making the final shape of the cover (Figure 11).

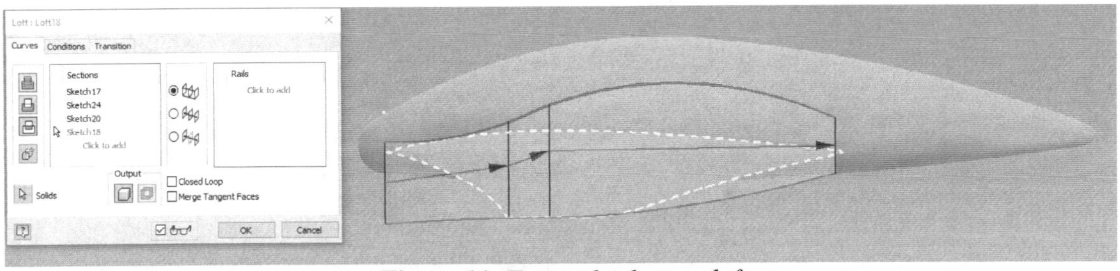

Figure 11. *Front wheel cover loft*

The Loft command should be set as to cut the extruded body, not the car chassis.

Then the front wheel cover should be mirrored at the y-z plane then the three pieces should be united. The rear wheel cover is made by a similar method.

The windscreen is made by the „Split" command. (It is only made for visual design because the cover is fully made by a transparent foil, which will be foiled then).

The result can be seen on the figures (Figure 12, Figure 13, Figure 14).

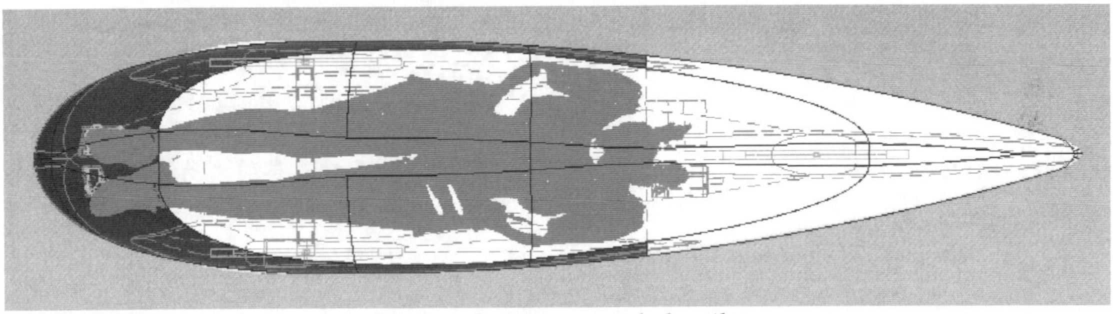

Figure 12. *Top view with the pilot*

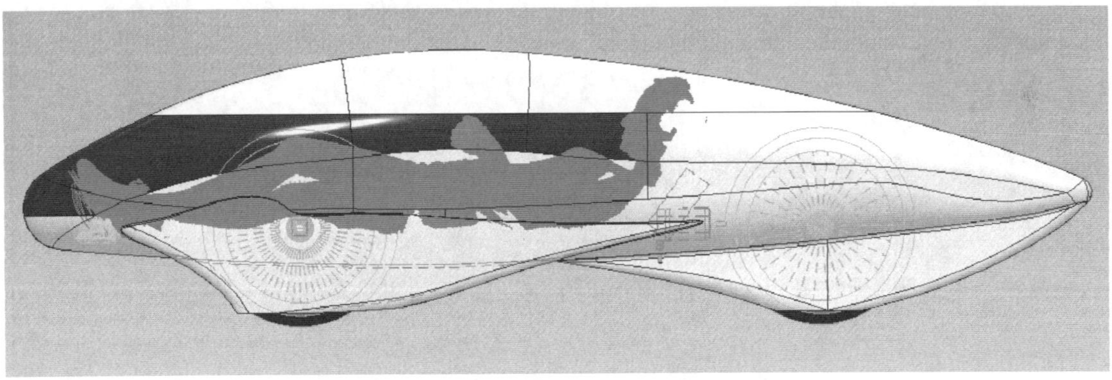

Figure 13. *Side view with the pilot*

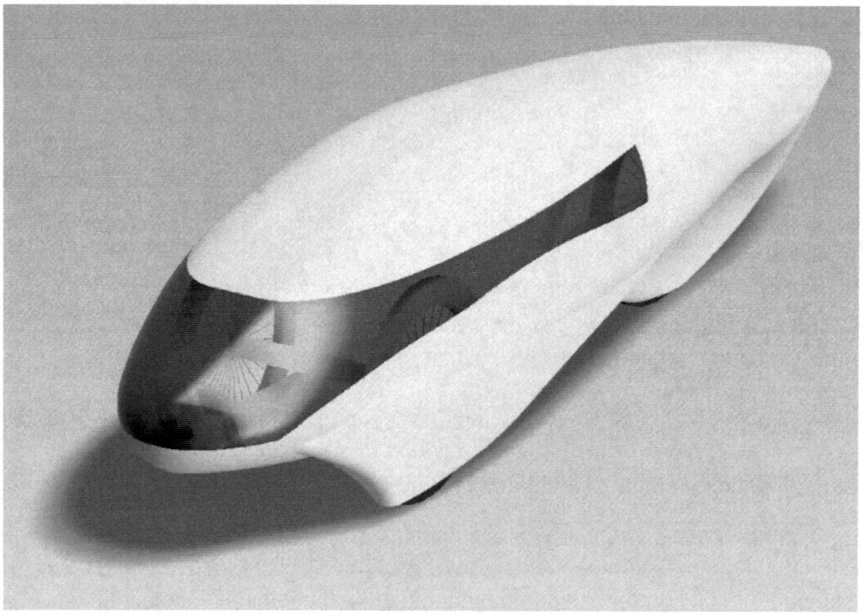

Figure 14. *STECO Team's 2022 racing car's visual design*

7. Air resistance

For a better result in the race, all of the unnecessary weight must be removed and the air resistance must be minimized.
Air resistance is important, because when the car begins to move the following happens.
Drag force affects the car with a vectors that are the opposite of the car's moving direction. The volume depends on the car's frontal area (A), velocity (v), drag coefficient (cw) and the density of air (ρ).

If the body in the fluid domain is not a perpendicular surface to the velocity vectors, then the drag coefficients correct the correlation (1).

$$F_e = \frac{\rho}{2} \cdot c_W \cdot A \cdot v^2 \quad (1)$$

The correlation from the left to right is:

F_e: drag force
$\frac{\rho}{2}$: half of the air density
c_W: drag coefficient
A: frontal surface of the vehicle
v: vehicle's velocity

Reducing the air resistance makes the car more efficient, the force needed for acceleration is lower and it can hold the needed velocity.

Air resistance is comprised of pressure and friction resistance.

The body's shape is made to look like a shape of a drop. In this case the friction resistance takes a bigger role in the full air resistance. [10]

7.1. The affection forces to the body

A fluid domain affects the solid body during the flow – it is braking it. This force is influenced and created by the pressure-shear voltage distribution on the body surface and the flow path's characteristics behind the vehicle.

A flow path is created behind every body inside the frictional loss-making fluid domain. At that place the pressure and velocity can be different to the undisturbed flow's characteristics. In general, a greater flow path behind, will cause more forces to affect the body. [10]

7.2. Calculating the drag coefficient

It is a comparison value for describing shape forms and a(n) unidimensional number for measuring the force of air resistance. [9] The lower drag coefficient causes a better vehicle shape according to air resistance.
The simulations about it are made by the Ansys Computational Fluid Dynamics (CFD) software. The models are made by the Autodesk Inventor Professional CAD modelling software.
The goal of the simulation is to determine the drag coefficient at 6.9 m/s velocity.
The software used during the simulation:

Ansys SpaceClaim
Ansys Fluent
Autodesk Flow Design

For the first step, the import of the file into SpaceClaim must done. Here I made the preparation of the model for the simulation. I made an enclosure around the body. Then I made groups about the velocity inlet, pressure outlet and the walls.
Then we did the meshing. The maximum size was 0.5, while the minimum was 0.05, with a poly-hexacore shape, with 10 layers (poly-hexacore shape is handled easier by the software and the 10 layers provide more precise calculations).

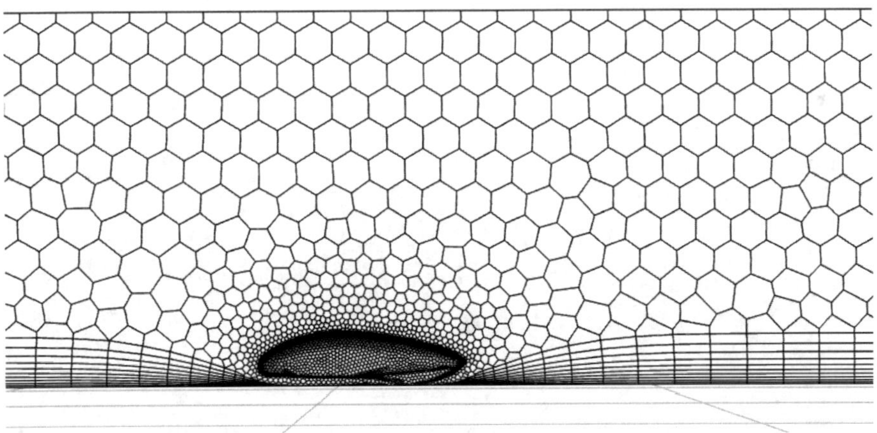

Figure 15. *Layers around and under the car*

For the solution, the parameters must be set up. Because of the body being streamlined, it is hard to define the laminar-turbulent flow's boundary. So, I used a(n) SST k-omega model. It is not resistant to these boundaries and the possible mesh failures. Then I set the magnitude of the velocity inlet. Other values were set that describe the car body, and I ran the simulation. For the drag coefficient we got 0.170.
On the body surface I determined the velocity magnitude as a result of the airflow. These values can be seen visual on the Figure 16.

Figure 16. *Air resistance's finite element analysis: flow velocities on the chassis surface*

The modification of the airflow is defined and displayed on the visual. These values show the changes of airflow, at the moment of the impact and the corresponding changes of the airflow. It also shows the car's wind-shade (Figure 17).

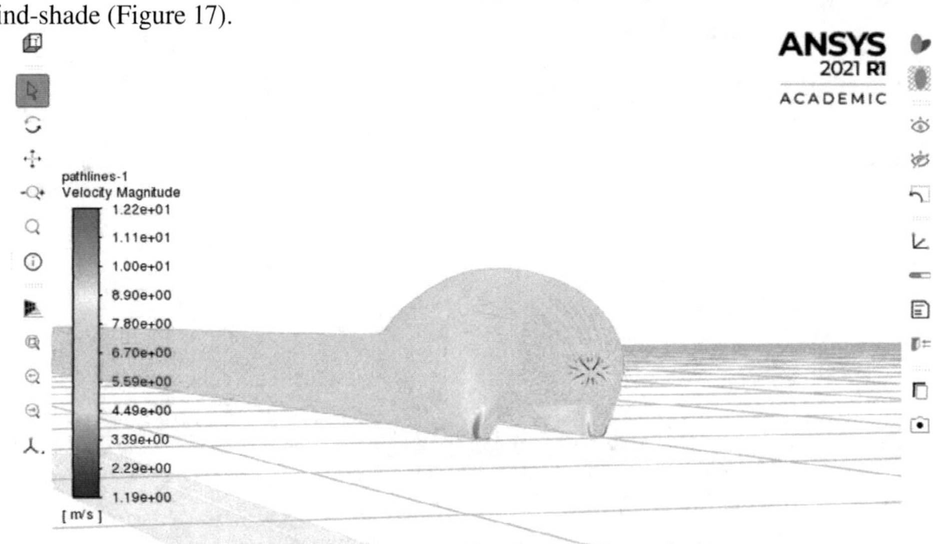

Figure 17. *Air resistance's finite element analysis: flow velocities on the chassis surface*

As a comparison I made a simulation using Autodesk Flow Design, with the new „Jupiter" car and our older „Zeus" Eco-Marathon car. Here the Jupiter's drag coefficient is 0.16. whereas Zeus' drag coefficient is 0.3 (Figure 18).

IOP Conf. Series: Materials Science and Engineering 1237 (2022) 012010 doi:10.1088/1757-899X/1237/1/012010

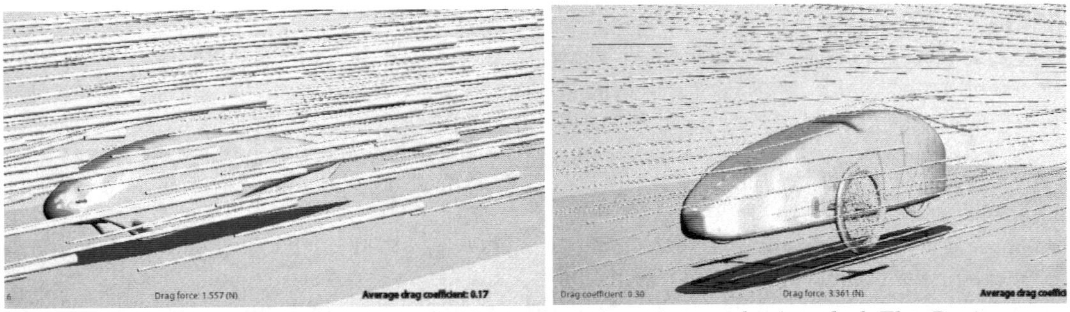

Figure 18. *Comparing the 2017 Zeus and 2022 Jupiter air resistance by Autodesk FlowDesign*

Table 1. The results of the air resistance simulations of the variants are shown in Figure 5

	„A" version („Zeus")	„B" version	„C" version	„D" version („Jupiter")
Drag force – F_{drag} [N]	3.361	0.574	0.579	0.497
Drag coefficient - c_w	0.3	0.196	0.199	0.170

The results of the study show that the air resistance of the new car is much more favorable.

8. Manufacturing technology

The body is made out of PETG plates using vacuum forming. The capacity of the forming machine cannot do the whole process for the entire body, so the chassis is divided to smaller pieces. The number of divided body sections is 10.

The working space of the machine is 1000x2000x2500 mm. Figure 16. shows the chassis division.

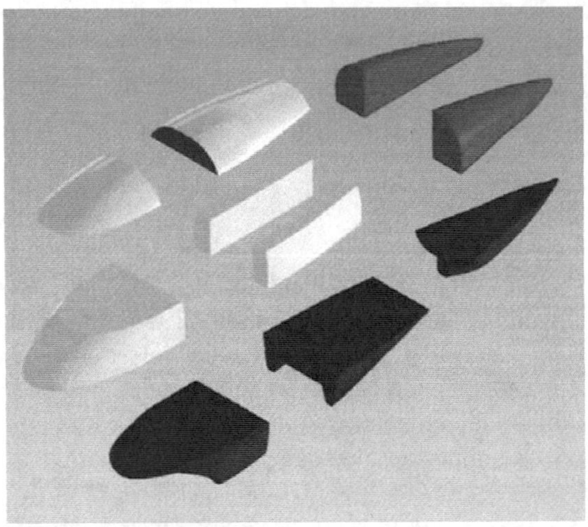

Figure 19. *Sections of the car chassis*

The model illustrates those solid bodies which will be on the forming desk. Different colors show the different areas of the chassis. White is the cockpit-door, red is the engine compartment door, blue is the bottom part that is fixed to the frame. The thin polycarbonate elements are assembled by riveting and attaching. The templates for the vacuum forming are made by MDF blocks and manufactured by CNC milling.

9. Frame

The chassis should be fixed to the frame via locking tabs, and the frame should be made according to this. The frame also provides the stiffness for the chassis. The vehicle's height is defined by the chassis' interior height.

After all the steering system and the wheel suspension must be made in the frame.

The front wheels' steering angle must be set up, as it must not touch any element of the frame or chassis.

In the engine compartment has to have enough space provided to accommodate the chain-drive, electric motor, battery and other control elements.

The cockpit is isolated from the engine compartment by an aluminum plate.

The frame (by derogation from the previous years) made by composite tubes and 3D printed connection nodes which are strengthened by glass-fiber.

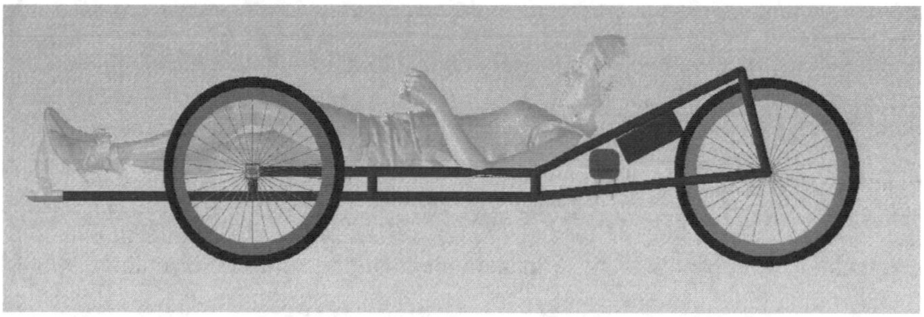

Figure 20. *Frame sideview*

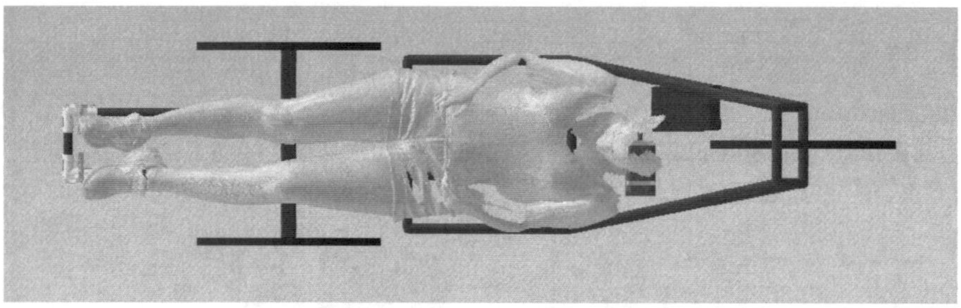

Figure 21. *Frame top view*

The frame's finite element analysis is done for acceleration, braking and turning loads.

10. Summary

During planning of the STECO team's new vehicle our main goal was to reduce weight and air resistance. For a lower total weight, our aluminum construction is replaced by composite carbon-fiber tubes. The connecting nodes are made by 3D printed glass-fiber strengthened composite. Another weight reduction technology is the making of the chassis. The elements are made by PETG plates with the technology of vacuum forming.

The shape of the car is another important part. The goal during the planning was the lowest possible air resistance. For this optimalization we examined three chassis versions. Using the results of an Ansys software, we chose the best version. In comparison the new chassis' air resistance is much lower than the previous versions and the previous car.

The wheels and tires of the vehicles are the same type so these rolling resistance remains the same. The total weight is much lower for the new car. The conclusion is that the Jupiter can take a longer trip with the same amount of energy!

Acknowledgements
This time we would like to thank our sponsors for supporting us. This costly project and the ability to enroll and take part in the competitions would not have been possible without them.

3i Fejlesztő és Szolgáltató Kft.
AVENTICS Hungary Kft., Emerson Automation Solutions
Corporeal.xyz Kft.
Derby-Túr Kft – Gázolajkút
Duna Aszfalt Kft.
eCon Engineering Kft.
Entra-Sys Kft.
FreeDee Printing Solutions
Havancsák-SKF Műszaki Szakáruház
Henkel Magyarország Kft.
Hirschmann Car Communication Kft.
Kvalix Automatika Kft.
Linamar Hungary Zrt.
Loctite
Maniac Kerékpárüzlet, Szerviz és Webshop
Medikémia Zrt.
Moltech AH Kft.
Princefibre-composite profiles
Schaeffler Savaria Kft.
SolvElectric Technologies Kft.
Spezial-Form Kft.
Szegedi Tudományegyetem Mérnöki Kar
SZTE Mérnöki Kar Kollégium
Tornádó International Kft.
Varinex informatikai Zrt.
VEKTOR Kft.

References

[1] https://www.makethefuture.shell/en-gb/shell-eco-marathon
[2] https://s04.static-shell.com/content/dam/royaldutchshell/documents/ecomarathon/results/sem-europe-2016-prototype-results-battery-electric.pdf
[3] https://www.shell.com/make-the-future/shell-ecomarathon/challenger-events/france/_jcr_content/par/textimage.stream/1498639679314/12b50662ffe073a67c680318c25166e654d59470/final-results-shell-eco-marathon-le-mans-2017.pdf
[4] https://www.shell.com/make-the-future/shell-ecomarathon/europe/results-and-awards/_jcr_content/par/expandablelist_781025105/expandablesection.stream/1563371146355/3ffc055693403502ab9323df9ff51f0dea80a2ef/europe-2019-prototype-battery-electric-full-results.pdf
[5] Tóth I.T.: „Drive alternatives. Is the future surely in the electric drive?" Proceedings of 3rd Agria Conference on Innovative Pneumatic Vehicles – ACIPV 2019. May 09, 2019. Eger. pp. 37-42. ISBN 978-963-449-149-1 https://pneumobil.hu/content/download/11552/45579/file/Proceedings%20of%203nd%20Agria%20Conference%20on%20Innovative%20Pneumatic%20Vehicles%20–%20ACIPV%202019.pdf

[6] Dr. Tóth I.T.: Renewable energy. Is it solution or the next huge business? International Conference on Science and Technique in the Agri-Food Business. Szeged -05-06.11.2008. ICoSTAF2008 ISBN978-963-482-908-9 pps. 392-397.
[7] https://www.makethefuture.shell/en-gb/shell-eco-marathon/global-rules
[8] Artec EVA specifikáció: https://cdn.artec3d.com/pdf/Artec3D-Eva.pdf
[9] Czuczor Mátyás: Teherautó mögött haladó személyautó ellenállás vizsgálata és a személygépjármú fogyasztására gyakorolt hatása, Miskolc-Egyetemváros, 2017. pp. 8., 12.
[10] Dr. Lajos Tamás: Az áramlástan alapjai, Budapesti Műszaki Egyetem, 1999. pp. 31.

Innovative methods and research directions in the field of logistics

T Juhászné Bíró[1] and P Németh[1]

[1]Department of Logistics and Forwarding, Széchenyi István University, Győr, GYMS, 9026, Hungary

biro.tamara@sze.hu

Abstract. By the 21st century, logistics and various supply chains had become key units in the global market and corporate structures. Industry 4.0 has brought developments and implementations to life that have drastically changed and are still changing the practices used in certain areas of logistics. Many new technologies (advanced robotics, additive manufacturing, artificial intelligence (AI), blockchain, drones, Internet of Things (IoT)) have emerged in the digital world, which many companies are using to develop cyber-physical systems in order to increase efficiency, speed, accuracy and the ability to change and steer competition between companies around the world. Planning tasks at the strategic, tactical and operational levels are covered in the areas of production and logistics. The tasks presented here can be identified as extremely complex optimization problems that belong to the np-hard complexity class. These can be addressed in many cases with metaheuristics, and industry also often uses search strategies inspired by biological or physical processes. Metaheuristic algorithms simulate the behavior of a selected phenomenon in a given search area. Algorithms based on various principles can help optimize processes, such as: population-based algorithms, evolutionary methods, behavior-inspired procedures, swarm intelligence methods, etc. New technologies or metaheuristic procedures are also increasingly used in logistics due to the complexity of the tasks. This paper presents theoretical application possibilities of digital transformation, AI and IoT in the field of logistics. The paper provides a further brief overview of the problems surrounding metaheuristics, supported by examples. The article shows the impact of different Industry 4.0 technologies on logistics. There is a shortage of such comprehensive studies, so the article helps provide insight into innovative optimization opportunities in a larger area - the field of logistics. Within this one paper, the impact of new technologies on the field of logistics was collected. A brief description of these will help to identify further directions and deepen the applicability of the new methods in logistics.

1. Introduction

In recent decades, the study and continuous development of various optimization processes has become a defining task and approach in almost every area of the industry. For companies, the rapid development of computing technology, the technologies brought to life by Industry 4.0, and a variety of intelligent solutions offer more and more opportunities to improve, among other things, speed, flexibility, and cost-effective solutions.

The benefits of the global market can best be reaped by companies setting up supply chains in different countries around the world. This process, of course, increases the complexity of the chains, making it a major challenge for partner companies. It is essential for the proper functioning of global integrations that modern information technology systems are used and utilized in logistics systems and supply chains. [1]

The Fourth Industrial Revolution (Industry 4.0) and the new technologies that have emerged with it have been the focus of attention over the past few decades. Researchers have also been attracted to new techniques (Internet of Things, advanced robotics, artificial intelligence, additive manufacturing, blockchain, drones). A number of studies have addressed their applicability, effects, scalability, and adaptations. Industry 4.0 technologies can also change best practices in logistics processes and supply chains. As a result, the research community of operational management will also have the opportunity to examine the new research issues raised by new technologies and the implications of their application. Under certain conditions, these new technologies have the potential to create economic, social or environmental value, and these conditions can be examined to maximize their value-creating impact on supply chains. [2]

The logistics industry also uses metaheuristics many times, as logistics problems belong to the np-hard complexity class due to their complexity, and one of the best solutions to solve them is these algorithms.

The paper provides a brief overview of Industry 4.0 technologies and their relationship to logistics and discusses the applications of metaheuristics and the problems surrounding them.

2. Technological Advancements in Industry 4.0

This chapter briefly introduces some of the Industry 4.0 technologies and their interfaces with logistics. Due to the diversity and complexity of logistics, numerous optimization tasks and problems arise in this area. Their solution has been a challenge in the past, but Industry 4.0 also offers opportunities to simplify decision-making processes in logistics, manufacturing, warehousing, shipping, or give new perspectives to individual areas or past practices.

One of the biggest demands of manufacturing companies as well as supply chains is to ensure their competitiveness in the global market. This also requires the key features of the Fourth Industrial Revolution: increased efficiency, productivity, flexibility and sustainability. Industry 4.0 technologies also support the development of an intelligent manufacturing environment system. These technologies include a number of methods (Oztemel and Gursev also refer to them as components of I4.0), which may include, but are not limited to: cloud-based systems, IoT platforms, 3D printing, big data analysis, algorithms, heuristics, metaheuristics, intelligent sensors, advanced robotics, etc. [4] [5]

2.1. Additive Manufacturing (AM)

Additive manufacturing (3D printing) is a revolutionary technology capable of producing objects with complex shapes and internal geometries for which conventional manufacturing methods are not suitable. Its huge advantage is that it does not require a new setting when switching between the production of objects of different shapes and geometries, thus eliminating the shortcomings of conventional manufacturing technologies - in many cases - in terms of flexibility, speed and cost. As a result, the speed of market entry will also increase significantly. These can also reduce the upfront production cost, as no separate production investment is required due to product specifications. It will be possible to fashion products produced in small quantities profitably. How does additive manufacturing affect logistics? AM may be able to change the entire supply chain architecture. It is a possible vision that for some products, instead of an operating strategy for mass production, which may be limited, your mass will be characterized by individual production. This change can be brought to life by AM's promised speed, quality, flexibility and cost benefits. [6]

Of course, many disadvantages currently limit the wide applicability of AM technologies: different materials require different AM technologies, size, strength, evenness of the product is also a bottleneck. Additionally, the cycle time of AM is usually slower than the cycle time of conventional manufacturing methods, so in some cases considerable time is required for post-processing. Because of these, AM currently performs better than conventional methods with lower production volumes. [2] [6]

2.2. The Internet of Things (IoT)

IoT is one of the latest IT developments, a network that consists of uniquely identifiable endpoints (or "things") that communicate over an IP connection without human intervention. Sensors at the end of the networks are connected to physical devices (such as trucks, freighters, wind turbines, drilling rigs, and smart meters). Efficiency in logistics processes according to Ellis et al. (2015) lies in: The amount of data released from these tools is related to supply chain processes. They provide unprecedented insight into real-time conditions when data is collected and analyzed efficiently and properly. This way, we can get an early warning of problematic situations and take immediate action on recovery. If the response to the signals is timely, the efficiency of the supply chain can rise to a new level. [7]

There are numerous definitions of IoT, many of which have been collected in the paper by Ben-Daya, Hassini, and Bahroun (2017), but in this study we consider what is proposed to define IoT in relation to supply chain management to be authoritative: „The Internet of Things is a network of physical objects that are digitally connected to sense, monitor and interact within a company and between the company and its supply chain enabling agility, visibility, tracking and information sharing to facilitatetimely planning, control and coordination of the supply chain processes." (Ben-Daya, Hassini and Bahroun, 2017) [8]

One of the biggest benefits of IoT is that companies can have accurate data on current inventory levels, including its environment. As a result, they are able to transport a much wider range and volume of data at both the production and logistics levels. Questions may arise as to what types of data are most useful and how best to use the data obtained. This can also be extremely useful information in other areas and for the application of other procedures (e.g. optimization algorithms). [6] Holdowsky et al. (2015) argue that one of the most significant capabilities of the IoT is its ability to generate information in previously unseen conditions where there is no human observation. [9] So IoT can really help to obtain information and data that can be of great help in real-time optimization through in-house but also other (devices, facilities) networks. This will allow to optimize at the system level and find the most effective solution to a particular problem. [10]

Verdouw et al. (2016), for example, suggest the use of IoT to manage the food supply chain, as Internet technologies enable supply chains to dynamically leverage virtualization in operational management processes as well. This virtualization also gives supply chain players the ability to monitor, manage, plan and optimize business processes over the Internet remotely and in real-time, and take immediate action in the event of discrepancies [11]

However, it should also be noted that IoT devices pose significant security and privacy issues (e.g., devices may disrupt other technologies, attack users' devices or systems, and some IoT devices may remain unattended, thereby exposing them to physical attacks), but the literature on possible solutions is fortunately growing year by year. [12]

2.3. The Blockchain

Companies want to store supply chain data sets more and more, thanks to IoT, in a secure but accessible way. The blockchain is a possible solution for this, which is a distributed and secure general ledger (stores information). The advantage is that it can be accessed, modified or supplemented from anywhere, the data is not stored in a central location, but on a so-called peer-to-peer network. It cannot be modified unilaterally after a new block has been added to the chain. Its distributed nature provides more flexibility than a central database and allows information about the entire history of a product to be stored as it travels throughout the supply chain. The information in the blockchain can be digitized, it can be automatically uploaded from the data of the IoT sensor, for example. The disadvantage may be that the quality of the information is inadequate, not factual, but in an IoT blockchain combination, if the sensor units are accurate, the information in the blockchain will be the same. An important application of the blockchain is also the availability of "smart contracts". This means that these contracts are automatically activated based on some externally controlled event (e.g. the container arrives at customs, payment can be authorized immediately). [6]

2.4. Artificial Intelligence (AI)

The term artificial intelligence nowadays means much more than human intelligence when it comes to the automation of thinking in terms of games (e.g., chess), proving mathematical theorems, making a diagnosis, or translating text. With a wealth of intelligent solutions, decision support software, optimization programs, and the rapid development of IT technologies, AI's algorithms are also capable of solving increasingly complex tasks, helping professionals work. Like Industry 4.0, AI does not have a single specific, exact definition, it is used in summary to indicate that it is not a person but a computer that is involved in problem solving. Machine learning, expert systems, neural networks, metaheuristics, IoT alone or together robotics, increase the efficiency and applicability of artificial intelligence. [6]

According to the paper by Vemuri et al. (2021), the application of AI contributes to the development of all undertakings. Its use also has an impact on the business environment, enhances the speed and productivity of business operations, raises the standard of monitoring, develops talent management skills, and expands the capabilities of the business organization. [13]

As a result of Industry 4.0 technologies, the use of digitization, IoT, and AI is becoming increasingly important in supply chains. The field of logistics is of great importance to the global market, which is why the term Logistics 4.0 summarizes the possibilities for optimizing logistics chains and networks using intelligent technologies. The goal of Logistics 4.0 is to adapt the flow of materials and information to the requirements of Industry 4.0 technologies, thereby increasing the efficiency of the entire supply chain. [14]

Toorajipour et al. (2021) carried out a comprehensive analysis of, inter alia, AI techniques in the field of logistics and supply chain. The following have been identified: field logistics: Automated planning, Artificial neural networks, Data mining, Agent-based/multi-agent systems, Simulated annealing, Robot programming, General forms of AI, Heuristics; field supply chain: Artificial neural networks, Agent-based/multi-agent systems, FL/modelling, General forms of AI, Physarum model, Swarm intelligence, Bayesian networks, Data mining, Stochastic simulation, Support vector machines. [15]

It can be seen, therefore, that AI's growing raison d'être in logistics and supply chains is gaining ground and emerging as a competitive advantage for its users.

2.5. Advanced Robotics

Manufacturing automation and robotics used in factories are already a widespread and increasingly sophisticated technology. Basically, they have been used to reduce the cost of expensive labor, improve quality (the robot will not get tired or become inattentive), or increase safety (e.g., the environment is dangerous to humans, the material, etc.). Due to its rapid development, it can have a major impact on factories, agricultural production, many services and even distribution logistics. The great advancement of sensor technology and artificial intelligence enables/can enable a new generation of robotic technologies that can work accurately, precisely and safely, and together with human workers. The latter raises a number of questions regarding the design and management of such systems. Its disadvantage is that its acquisition and installation can be time-consuming and can only be carried out by suitable, highly qualified professionals. Therefore, in many cases, installation can be expensive, costing several times the purchase price of the robot.

The latest and perhaps most exciting development in robotics - where there are also major advances - is the automation of vehicles or the increasing use of drones. For example, they can be used to survey remote locations, monitor certain devices, spray crops, or deliver smaller products. Robotic delivery options can have a significant impact on supply chain design, however this still requires the different robot technologies to improve their cost, flexibility, speed, or even quality. [6]

3. Metaheuristics

Key corporate goals include continuously improving optimization procedures to reduce computation times, inventory costs, and lead times, increase customer satisfaction, and enable companies to make higher profits. In the areas of supply chains as well as production and logistics, metaheuristic procedures are increasingly used due to the complexity of processes and tasks.

Countless metaheuristic algorithms are noted in the literature, and more and more new methods are emerging in this field every year. Not to mention that for many problems, the best solution can be achieved by using multiple metaheuristics together (hybrid versions), so the number of methods used is much larger than the existing single algorithms. [3] [16]

Metaheuristics can be grouped according to several aspects, Talbi (2009) and Liao and Li (2020) similarly divided them into single-solution based metaheuristics (e.g. Iterated Local Search, Greedy Randomized Adaptive Search Procedure, Local Search, Simulated Annealing, Tabu Search, etc.) and population-based metaheuristics (e.g., Evolutionary Algorithms, Swarm Intelligence, Artificial Immune System, etc.). [17] [18] According to Maier et al. (2019) population-based evolutionary algorithms are one of the main and most popular classes of metaheuristics. [19]

There are numerous advantages to using metaheuristics, but according to Swan et. al, one of the biggest advantages is that „the underlying search logic can be applied to any problem which can be decomposed into a few elementary aspects, namely solution representation, solution quality evaluation and some notion of locality." (Swan et. al, 2022, p.394) [16]

3.1. Problems surrounding the discipline of metaheuristics

As part of my doctoral research, I have examined a number of metaheuristic algorithms (on a theoretical level), especially those that have already been used successfully in the field of logistics. My hypothesis is that a logical connection can be discovered between the given problem, the input data, and the applied procedure, and from these a framework methodology can be created to associate the algorithms and the tasks to be solved. However, even after reading plenty of articles, it is difficult to discover logical connections and possible similarities, even though there are thousands of researches on this topic. The scientific community has also recognized that there are numerous problems in this area, such as:

- there is a lack of summary studies that demonstrate the common features of metaheuristics,
- in many cases, researchers ignore existing studies, published procedures, and present in their own writing even algorithms that have been tested for similar problems in the past,
- in many cases a study does not clearly describe why the given algorithm works best, what are the framework conditions, constraints, etc.
- source codes are most often not available and other data is missing for adaptations. [16]

The problems described can often lead to "double discovery" and stand in the way of the applicability and evolution of metaheuristics. To address these, a research initiative called "Metaheuristics in the Large" (MitL) has been set up to reduce problems in this field, both theoretically and practically. A further aim is for scientists to use existing procedures and ideas in the literature over and over again and even to communicate with each other about the information needed for adaptations. [16]

Understanding metaheuristic methods is also very important because industrial tasks are becoming more and more complex and in many cases these algorithms are suitable for solving them. The procedures used could be used more widely and in more areas with the right knowledge. However, this requires the elimination or reduction of shortcomings, to which I would also like to contribute with my research.

3.2. Examples of the application of metaheuristics in the field of logistics

In the field of logistics, metaheuristic procedures are also increasingly used. Numerous new procedures and algorithms are appearing for even better optimization, as well as the involvement of logistics sub-areas where the use of metaheuristics was not typical before. In the following, we provide a brief overview of these.

3.2.1. Solving scheduling problems with metaheuristics. The complexity of scheduling problems is constantly increasing due to the increasing complexity of supply chains. In the study of Coelho and Silva

(2021), the metaheuristics already applied to different shop configurations (Job Shop, Flexible Job Shop, Flow Shop, Flexible Flow Shop) were collected. Fifteen different algorithms have been identified, some of which are: Genetic Algorithm, Bat Algorithm, Taboo Search, Variable Neighborhood Search, Hybrid Simulating Annealing-Genetic Algorithm, Hybrid Tabu Search-Particle Swarm Optimization, etc. This article also supports the theory mentioned earlier, according to which comprehensive collections dealing with metaheuristic procedures in a given field or a given problem are lacking. It can be seen that many researchers have already dealt with the metaheuristic procedures used for shop configurations, and several have described the same algorithm or hybrid versions. Based on these, the question may legitimately arise as to whether completely new studies, comparative analyses were needed, or whether the applicability of the given procedure would have been "merely" expanded with an appropriate adaptation. [4]

Another common logistics scheduling problem is the resource-constrained project scheduling problem (RCPSP), which has a wide variety of applications in many areas. A review study by Pellerin, Perrier, and Berthaut (2020) identified 27 metaheuristics applied or suggested for RCPSP, which were classified into three major classes [20]:

- Local search metaheuristics (for example: Greedy randomized adaptive search procedure, Local search, Taboo Search, etc.),
- Population-based metaheuristics (for example: Ant colony optimization, Firefly algorithm, Genetic algorithm, Memetic algorithm, Particle swarm optimization, Scatter search, etc.),
- Learning metaheuristics (Neural network). [20]

It can be seen as well, the examination of the myriad metaheuristics for a given problem, but there is a lack of a concise, conclusive study to support further developments and adaptations.

3.2.2. Recent metaheuristics of decision-making processes. Making a decision is a difficult and complex task, as there are many factors to consider. Automating decision-making processes is particularly complex, as in the traditional case, human experience and intuition help the process. Therefore, in many cases, decision support software works well to help people with these tasks. However, there is a lot of data that can be parameterized, so the decision-making process can also be described as an optimization problem. In this case, metaheuristics may be suitable for this type of problem solving, and ant colony optimization, genetic algorithms, or particle range optimization have already been used in decision-making processes. Metaheuristic algorithms can be combined not only with each other, but also with other types of methods. For example, the study by Keivanian and Chiong (2021) presents a fuzzy-metaheuristic approach in which the benefits of each method can be used to achieve even better results in a hybrid approach. The method is based on the Imperialist Competitive Algorithm (ICA), which is one of the latest metaheuristics and has already successfully solved a number of complex, practical operations. Improved versions of this (FAEICA: fuzzy adaptive and enhanced ICA and MOFAEICA: multi-objective FAEICA) and their theoretical success in Multimodal Single and Multi-Objective Optimization Problems are detailed in the paper of Keivanian and Chiong (2021). [21]

3.2.3. Additional new metaheuristics. Metaheuristic algorithms are methods that provide robust performance, adaptability, and flexibility to solve optimization problems. In practice, the tasks of algorithms can be divided into 2 major search procedures: the first is exploration, which operates in the whole area, and the second is exploitation, which examines a local part of the search space. [22] Algorithms of this type usually "work" on the basis of their food-seeking, navigational, or collective intelligence behaviors and simulate these behaviors to solve a given problem. The introduction of new metaheuristics to solve various search problems is growing year by year and is leading to the discovery of many new algorithms. In addition, in recent years there has been an accelerating pace in the introduction of new procedures. The following are some of the new metaheuristics that were created or introduced between 2014 and 2021: Jellyfish Search, Galactic Swarm Optimization, Whale Optimization Algorithm, African Vultures Optimization Algorithm, Honey Badger Algorithm etc. [23] [24]

The number of metaheuristic algorithms is constantly expanding and this will be a definite trend in the coming years. However, the pairing and co-application of the processes will also increase drastically, as the benefits of each process can be best exploited in hybrid processes. [25]

4. Conclusions and future work

The aim of our paper is to provide an overview of Industry 4.0 technologies and metaheuristic algorithms, especially in the field of logistics. We have discussed the pros, cons, and problematic issues of these technologies.

Industry 4.0 technologies have a direct impact on both logistics and supply chains: they can change and influence their structure, architecture, past best practices and processes. All this, of course, in order to create tasks and optimal solutions more efficiently, flexibly, faster, or even cheaper. This raises further research directions, as the benefits, disadvantages, and limitations of the Fourth Industrial Revolution need to be understood, if Industrial 4.0 technologies are to be put to good use in practice.

In the field of metaheuristics, an accurate understanding of procedures is important, and more information is needed for practical application and adaptation of each algorithm to another area or problem. This is not an easy task and, as we highlighted in the article, there are a number of problems surrounding the field of metaheuristics. Nevertheless, the use of metaheuristic algorithms offers a good solution for tasks belonging to the np-hard complexity class, even though in many cases these methods do not guarantee an accurate and optimal solution but are generally able to provide an acceptable approximation. A further question arises as to their industrial applicability for solving real engineering tasks. This is perhaps the biggest challenge, especially for recently released algorithms. Several studies deal with the evaluation of algorithms on design-engineering problems and the test results are generally encouraging. [26] [27] [28] [29]

My goal is to identify additional logistics areas where metaheuristic procedures may have a raison d'être. Furthermore, I am constantly examining the algorithms in order to explore further opportunities and gaps in their application.

References

[1] Gunasekaran A and Ngai E W T 2014 Expert systems and artificial intelligence in the 21st century logistics and supply chain management *Expert Systems with Applications* **41** 1-4

[2] Tang C S and Veelenturf L P 2019 The strategic role of logistics in the industry 4.0 era *Transportation Research* E **129** 1-11

[3] Lang S, Schenk M and Reggelin T 2019 Towards Learning- and Knowledge-Based Methods of Artificial Intelligence for Short-Term Operative Planning Tasks in Production and Logistics: Research Idea and Framework *IFAC (International Federation of Automatic Control) PapersOnLine* **52-13** 2716–21

[4] Coelho P and Silva C 2021 Parallel Metaheuristics for Shop Scheduling: enabling Industry 4.0 *Procedia Computer Science* **180** 778–786

[5] Oztemel E and Gursev S 2020 Literature review of Industry 4.0 and related technologies *Journal of Intelligent Manufacturing* **31** 127–182

[6] Olsen T L and Tomlin B 2019 Industry 4.0: Opportunities and Challenges for Operations Management *Manufacturing & Service Operations Management* **22(1)** 113-122

[7] Ellis S, Morris H D and Santagate J 2015 IoT-Enabled Analytic Applications Revolutionize Supply Chain Planning and Execution *International Data Corporation (IDC) White Paper* (www.idc.com)

[8] Ben-Daya M, Hassini E and Bahroun Z 2017 Internet of things and supply chain management: a literature review *International Journal of Production Research*, **57** 4719-42

[9] Holdowsky J, Mahto M, Raynor M E and Cotteleer M 2015 *Inside the internet of things: A primer on the technologies building the IoT* Deloitte University Press

[10] Evans P C and Annunziata M 2012 Industrial internet: Pushing the boundaries of minds and machines *General Elektric*

[11] Verdouw C N, Wolfert J, Beulens A J M and Rialland A 2016 Virtualization of food supply chains with the internet of things *Journal of Food Engineering* **176** 128-136

[12] Koohang A, Sargent C S, Nord J H and Paliszkiewicz J 2022 Internet of Things (IoT): From awareness to continued use *International Journal of Information Management* **62** 102442

[13] Vemuri V P, Priya, Naik V R, Chaudhary V, RameshBabu K and Mengstie M 2021 Analyzing the use of internet of things (IoT) in artificial intelligence and its impact on business environment *Materials Today: Proceedings* (Article in press)

[14] Oleśków-Szłapka J, Wojciechowski H, Domański R and Pawłowski G 2019 Logistics 4.0 Maturity Levels Assessed Based on GDM (Grey Decision Model) and Artificial Intelligence in Logistics 4.0 - Trends and Future Perspective *Procedia Manufacturing* **39** 1734–42

[15] Toorajipour R, Sohrabpour V, Nazarpour A, Oghazi P and Fischl M 2021 Artificial intelligence in supply chain management: A systematic literature review *Journal of Business Research* **122** 502-517

[16] Swan J, Adriaensen S, Brownlee A E I, Hammond K, Johnson C G, Kheiri A, Krawiec F, Merelo J J, Minku L L, Özcan E, Pappa G L, García-Sánchez P, Sörensen K, Voß S, Wagner M and White D R 2022 Metaheuristics “In the Large” *European Journal of Operational Research* **297** 393-406

[17] Talbi E G 2009 *Metaheuristics: From design to implementation* John Wiley and Sons, Inc., Hoboken, New Jersey

[18] Liao T W and Li G 2020 Metaheuristic-based inverse design of materials – A survey *Journal of Materiomics* **6** 414-430

[19] Maier H R, Razavi S, Kapelan Z, Matott L S, Kasprzyk J, and Tolson B A 2019 Introductory overview: Optimization using evolutionary algorithms and other metaheuristics *Environmental Modelling and Software* **114** 195-213

[20] Pellerin R, Perrier N and Berthaut F 2020 A survey of hybrid metaheuristics for the resource-constrained project scheduling problem *European Journal of Operational Research* **280** 395-416

[21] Keivanian F and Chiong R 2021 A Novel Hybrid Fuzzy–Metaheuristic Approach for Multimodal Single and Multi-Objective Optimization Problems *Expert Systems with Applications* (Available online 16 November 2021, 116199)

[22] Osuna-Enciso V, Cuevas E and Castañeda B M 2022 A diversity metric for population-based metaheuristic algorithms *Information Sciences* **586** 192–208

[23] Abdollahzadeh B, Gharehchopogh F S and Mirjalili S 2021 African vultures optimization algorithm: A new nature-inspired metaheuristic algorithm for global optimization problems *Computers & Industrial Engineering* **158** 107408

[24] Hashim F A, Houssein E H, Hussain K, Mabrouk M S and Al-Atabany W 2022 Honey Badger Algorithm: New metaheuristic algorithm for solving optimization problems *Mathematics and Computers in Simulation* **192** 84–110

[25] Seyyedabbasi A, Aliyev R, Kiani F, Gulle M U, Basyildiz H and Shah M A 2021 Hybrid algorithms based on combining reinforcement learning and metaheuristic methods to solve global optimization problems *Knowledge-Based Systems* **223** 107044

[26] Gupta S, Abderazek H, Yıldız B S, Yildiz A R, Mirjalili S, and Sait S M 2021 Comparison of metaheuristic optimization algorithms for solving constrained mechanical design optimization problems *Expert Systems With Applications* **183** 115351

[27] Bottani E, Casella G and Murino T 2021 A hybrid metaheuristic routing algorithm for low-level picker-to-part systems *Computers & Industrial Engineering* **160** 107540

[28] Cano J A, Espinal A C and Montoya R G 2020 Using genetic algorithms for order batching in multi-parallel-aisle picker-to-parts systems *International Journal of Applied Decision Sciences* **13(4)** 417-434

[29] Cortés P, Gómez-Montoya R A, Muñuzuri J and Correa-Espinal A 2017 A tabu search approach to solving the picking routing problem for large- and medium-size distribution centres

considering the availability of inventory and K heterogeneous material handling equipment *Applied Soft Computing* **53** 61–73

Software developments for an electric motor test bench developed at the Faculty of Engineering of the University of Debrecen

É Ádámkó[1], A Szántó[2] and G Á Sziki[3]

[1,2,3]University of Debrecen, Faculty of Engineering, 4028 Debrecen, Ótemető street 2-4. Hungary

adamko.eva@eng.unideb.hu

Abstract. At the Faculty of Engineering of the University of Debrecen, our research group has long-term experience in the dynamic modeling and simulation of electric motors. Several simulation modules have been developed during the last few years. The input data of the developed simulation modules are the technical parameters and characteristics of the analyzed motors. The mentioned technical data from the manufacturers is either not provided or it's incomplete, thus, it is necessary to measure it. Additionally, the accuracy of the output dynamic functions, generated by the simulation modules, has to be tested experimentally. As a consequence, it was decided by the research group to realize a Measurement System for performing both above-mentioned experimental tasks routinely, at high accuracy and at a precision level. In this paper, a brief description of the Hardware Component of the Measurement System and a full detailed one of the developed Software Component is given. The Software Component of the Measurement System is implemented in NI LabVIEW, Java SE, and VBA with an underlying MySQL database. Applying the Software Component, the measured quantities can be monitored in real-time, displayed graphically while being stored in a database for later use.

1. Introduction

Our research group at the Faculty of Engineering of the University of Debrecen has long-term experience in the modeling and simulation of electric motors and powertrains [1] [2] [3] [4]. To be able to simulate the above motors, their electromagnetic and dynamic characteristics have to be precisely known, since these data serve as input of the simulation program. Most of these data are not given in the catalog – specification – of the motor, so they have to be measured. These missing parameters was the main motivation to develop a complex Measurement System (MS) for the experimental study and testing of electric motors. By applying the MS, all of the electromagnetic and dynamic characteristics of a motor can be measured, additionally, test measurements can be performed on it too [5] [6]. The MS consists of a Hardware Component (HC) and a Software Component (SC). In Figure 1. that particular part of the HC is shown, which is used to perform test measurements on electric motors.

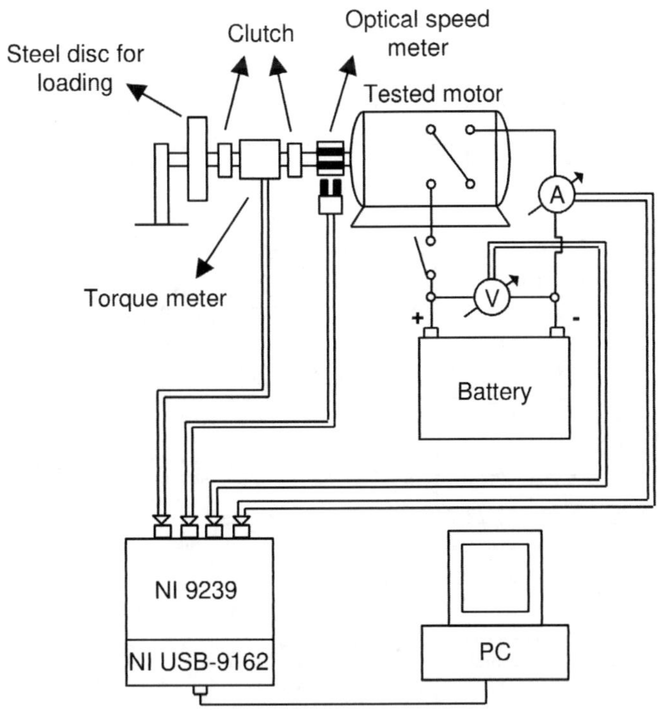

Figure 1. Hardware Component of the Measurement System for test measurements on electric motors

In the course of a typical test measurement, additional loads are applied to the motor with different moment of inertias. As voltage is switched on the motor, its rotor spins up from a state of rest until it reaches the maximum speed possible. During spin up, the angular speed and torque of the motor are measured in real-time. Additionally, the intensity of the electric current flowing through the windings of the motor, and the terminal voltage of the battery is measured, also in real-time. Several sensors are installed into the MS, e.g., optical LED and rotary torque sensors. The output signals of the optical LED and rotary torque sensors – for angular speed and torque measurements –, and also the intensity of the electric current, flowing through the motor, are converted to voltage signals. As it is shown in Figure 1. the output voltage signals of the different sensors and resistors are fed into a(n) *NI 9239* voltage input module, which is connected to a PC through a *CompactDAQ Chassis*. To be able to collect, separate, display, and analyze the digital data transmitted by the connector, a suitable software is necessary. That is the reason for the decision, why a self-developed SC was realized. The SC is primarily implemented in the NI LabVIEW environment [7][8]. In Section 2 the HC is briefly presented, and in Section 3 a detailed description of the SC of the MS is given.

2. Description of the Hardware Component

The wiring diagram of that particular part of the HC which is used during the test measurements on electric motors is shown in Figure 2. As it is mentioned earlier the output voltage signals of the different sensors and resistors are fed into a(n) *NI 9239* voltage input module, which is connected to a PC through a *CompactDAQ Chassis*.

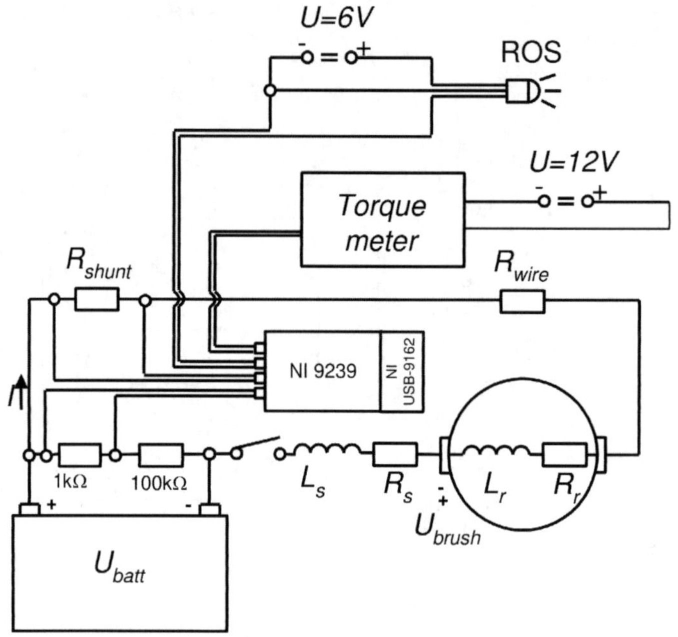

Figure 2. Wiring diagram of the Hardware Component of the Measurement System

The angular speed of the rotor is measured by an optical led sensor – *Model: ROS-P, Monarch Instrument.* The laser light emitted by the instrument is reflected by small foils fixed on the circumference of the rotor at equal distances. The reflected light then hits the sensor inside the instrument which generates a voltage signal in its output. The height of the output voltage signal is equal to the supply voltage of the sensor ($U_{OLS} = 6\ V$). A typical output voltage signal of the optical led sensor vs. time can be seen in Figure 3.

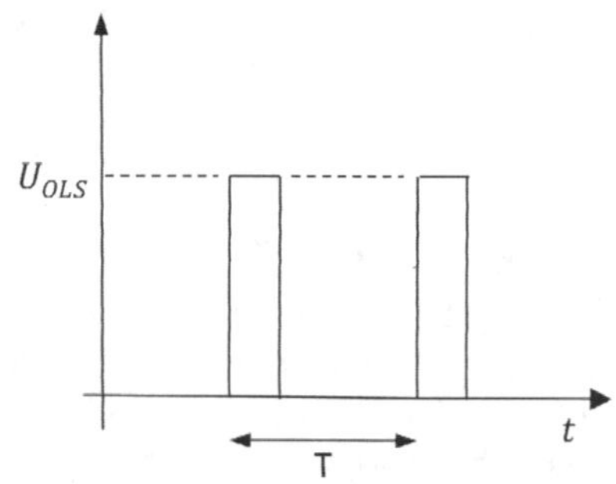

Figure 3. An output voltage signal of the optical led sensor vs. time

The time gap between two neighboring signals is calculated as:

$$T = \frac{60}{n \cdot N} \quad (1)$$

Where n is the RPM of the motor, and N is the number of the light reflecting foils fixed on the rotor. The torque of the motor is measured by a rotary torque transducer – *HBM, T22/200 Nm* –, with an accuracy class of 0.5. The applied torque on this transducer is linearly proportional to its output voltage (U_{RTD}):

$$M = C \cdot U_{RTD} \quad (2)$$

The value of constant C is equal to 40 Nm/V. The intensity of the electric current flowing through the windings of the motor is measured as a voltage drop on a shunt resistor with an electric resistance (R_{SHUNT}). Thus, the intensity of the electric current is calculated from the measured output voltage as:

$$I = \frac{U_{SHUNT}}{R_{SHUNT}} \quad (3)$$

The terminal voltage of the battery – since it is usually bigger than the maximum allowed input voltage of the *NI 9239* module –10 V – is measured by applying a voltage divider with a ratio of M : N, typically 1:100. The connection between the output voltage measured on the divider (U_{DIV}) and the terminal voltage of the battery is:

$$U_{BATTERY} = \frac{M+N}{M} \cdot U_{DIV} \quad (4)$$

3. Description of the Software Component

As it is explained above, the MS consists of two parts the HC and the SC. The main purposes of the SC are to separate, collect, store, process, analyze and graphically display the measured and the calculated data. In Figure 4. the structure of the SC is presented.

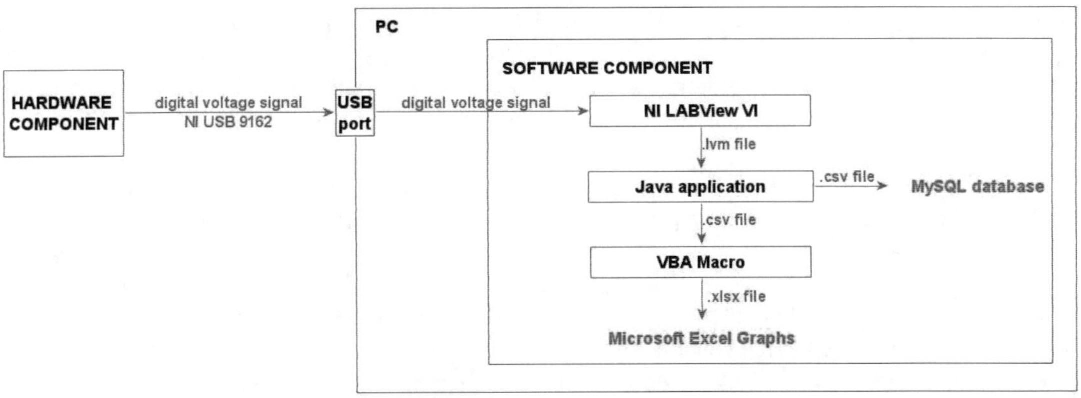

Figure 4. Structure of the Software Component

The SC contains four different modules; the first one was implemented in NI LabVIEW and provides the collection, separation, and text file output of the measured data that is coming from the sensors. Furthermore, it is capable of displaying both the raw voltage signals graphically, and the calculated output values numerically in real-time. The Front Panel of the VI serves as a User Interface for the whole SC. The second module is a command-line application that was implemented in Java SE [9], with the help of the Apache NetBeans IDE. The Java application provides the tool to pre-process aka clean and filter the raw data collected by the NI LabVIEW VI, and to make the connection – insert and retrieve data – with the underlying MySQL database. The third module is a VBA Macro, which was created for Microsoft Excel, which is the most frequently used software by engineers, although it is not commonly referred to as an engineering software by definition. Excel itself has limitations when it comes to processing a huge amount of data. That is why it was necessary to insert the command-line Java application into the process flow. For example, it took more than ten minutes (that was the upper limit of the developer's patience) for Microsoft Excel to filter an NI LabVIEW measurement file – which consists of almost three hundred thousand lines; while the Java command-line application solved the same problem in 1 second. The VBA Macro generates the desired output values and visualizes them, by using Microsoft Excel charts. The fourth module is the MySQL database, which is a simple OLTP database, with two tables to provide a long-term storage place for the measured data.

3.1. NI LabVIEW VI

Four analog input channels are used for collecting analog voltage signals by the NI 9239 measurement card. The channels collect signals from different sensors which are then converted to a digital signal by the built-in analog-digital converter of the card. The input circuitry of the measurement card can be seen in Figure 5.

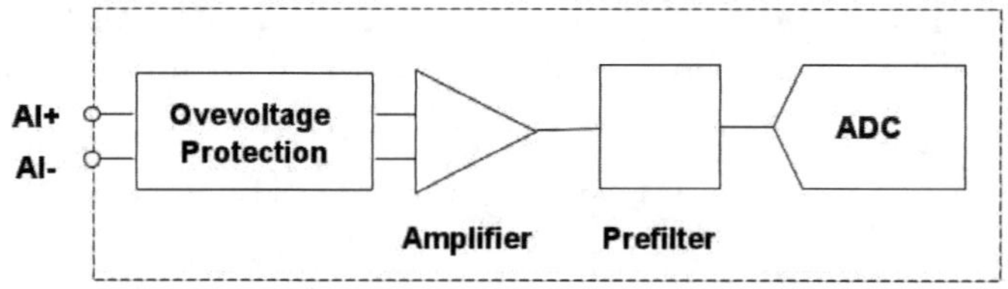

Figure 5. The input circuitry of the NI 9239

The resulting digital voltage signal serves as an input for the NI LabVIEW VI. To be able to process the input, first it is necessary to insert a DAQ I\O control for the physical channel in the Block Diagram of the VI. Several solutions are available in the NI LabVIEW to retrieve data from a physical channel, and in the NI DAQmx collection as well. For faster processing, the step-by-step solution was chosen by the developer instead of the DAQAssistant; which provides an off-the-shelf, but much slower solution. The step-by-step solution gives more freedom when it comes to deciding how to process the input data. Typical steps of data processing with the DAQmx VI-s can be seen in Figure 6. The displayed data flow reads from a physical channel which transmits a voltage signal on a single channel. The processing steps are the following according to Figure 6.:

- Firstly, a virtual channel is made for acquiring voltage input from the physical channel.
- Secondly, the properties of the reading process are set by a timing VI. The sampling rate and mode must be given here.
- Thirdly, – before reading – a data acquisition task is started.
- Fourthly, the suitable read VI – number of channels and number of samples fit to the physical channel – is called to read data with different purposes.
- Fifthly, methods and operations are executed to meet the purpose of the VI. In that particular example, the measured data is written into a(n) NI LabVIEW measurement file.
- Sixthly, when the reading is finished the task has to be stopped and cleared.

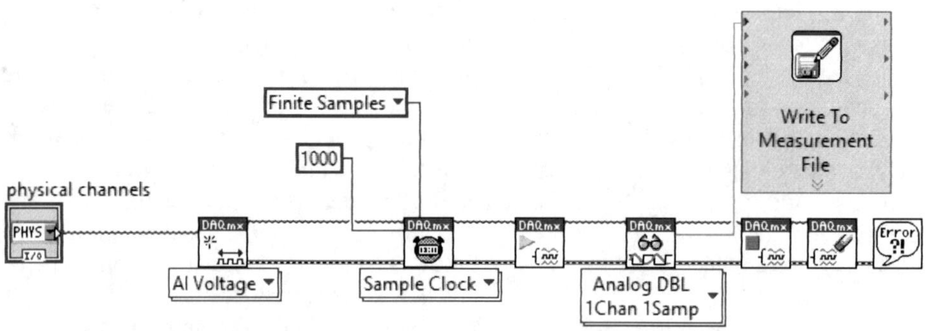

Figure 6. Typical data process flow with the DAQmx VI collection (voltage signal, single-channel, finite sampling mode and 1000 sample rate)

In our case, not one, but four voltage signals have to be processed because the measurement card has four analog inputs. Thus, instead of using a one-channel one-sample type of read VI, a multi-channel multi-sample type was used. The NI LabVIEW VI is capable of separating the data arriving from the four different output channels and displaying them on waveform charts versus the number of samples in the actual loop. Also, by applying some data manipulation methods and mathematical operations – with predefined input parameters, set on the User Interface, – the voltage, current, RPM, and torque output values can be calculated and displayed in indicators on the Front Panel of the self-developed VI. The VI provides collect and store functionality, by writing the measured data into the NI LabVIEW measurement file.

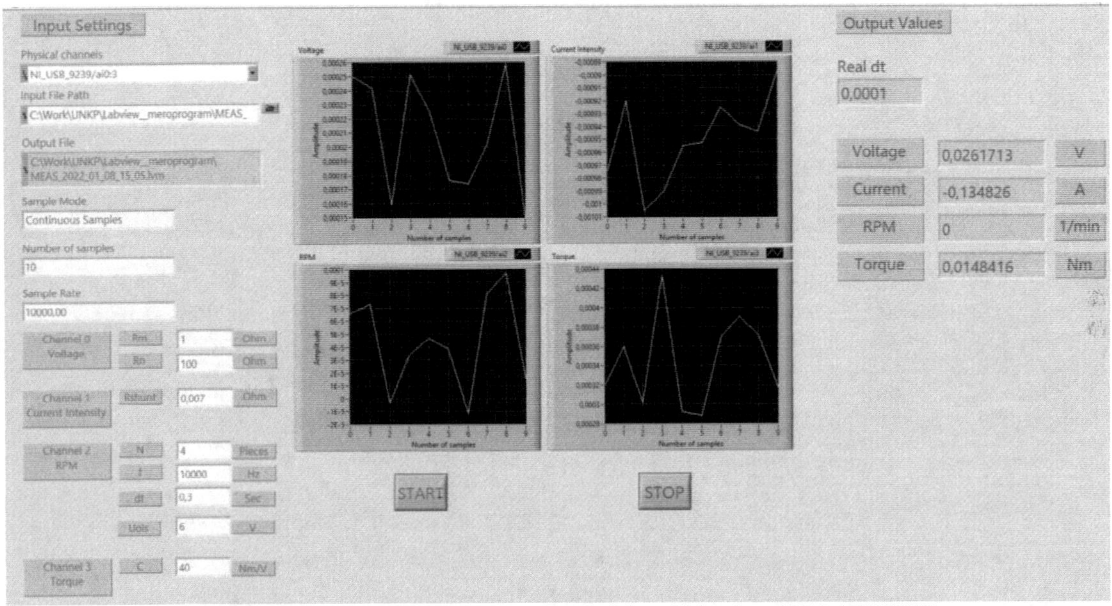

Figure 7. User Interface aka Front Panel of the self-developed NI LabVIEW VI

3.2. *Java command-line application*

The output data of the above-mentioned NI LabVIEW VI is one or more NI LabVIEW measurement files, which – according to the length of the measurement – can contain a few, but mostly a huge quantity of measured data in rows. Every row contains a time column and one to the four measured value columns, based on the number of input channels used. The developed Java command-line application inputs are the aforesaid NI LabVIEW measurement files (.lvm). Two main tasks have to be realized by the java application:

- The first is to insert measured values from all of the row input found in the NI LabVIEW measurement file/files into the underlying MySQL database, thus realizing the store feature of the SC.
- The second is to filter the rows based on the time gap between two measurements (dt); given by the user on the User Interface of the SC, to pre-process the data.

In default, measurements are recorded every 0.0001 seconds, by the self-developed VI. By applying a filter in the time gap values, a much smaller file size can be achieved. The third step of the process method is analyzing the data with Microsoft Excel. As it is mentioned earlier above, the application cannot deal easily with too many lines. The Java command-line application converts the input NI LabVIEW measurement files into 'comma-delimited' text files (.csv) because of compatibility issues.

3.3. VBA Macro

Microsoft Excel is easily automatable with the help of macros written in Visual Basic for Applications. It is a high-level, object-oriented programming language. The main task of the developed VBA Macro is to **analyze and display the processed results graphically** in charts. When the "comma-separated values" text file is opened in Microsoft Excel, the macro converts it into a Microsoft Excel workbook (.xlsx). Then, based on the input parameters given in the User Interface of the SC, all of the calculations – explained previously in Section 2 – are done, to gain the desired output values, like the voltage, current intensity, RPM, and/or the torque. As the last step, one or more graphs are built and displayed to be viewed later for investigation of the output values from the experimental tests. In Figure 8. one output Microsoft Excel workbook is presented, which was made during an experiment where the RPM needed to be calculated.

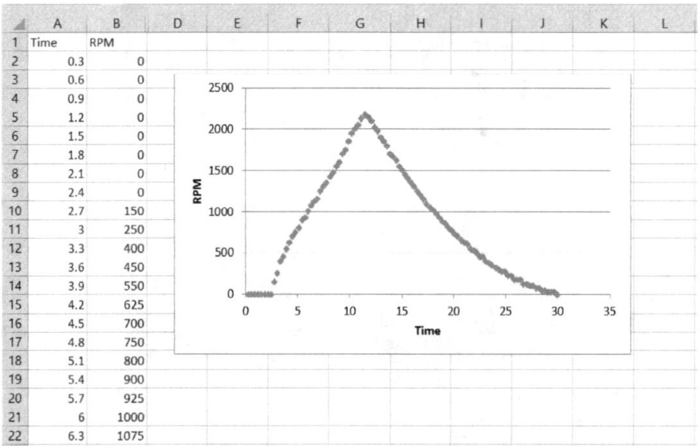

Time	RPM
0.3	0
0.6	0
0.9	0
1.2	0
1.5	0
1.8	0
2.1	0
2.4	0
2.7	150
3	250
3.3	400
3.6	450
3.9	550
4.2	625
4.5	700
4.8	750
5.1	800
5.4	900
5.7	925
6	1000
6.3	1075

Figure 8. Results of test measurement for RPM with the self-developed VBA Macro

3.4. MySQL database

The MySQL database serves as a long-term storage place for the measured data, collected during different experimental test measurements. The collected data can provide a solid base for statistical research about the investigated motors, or the properties and behavior of the applied sensors as well. It is a relational database that contains two tables connected by a foreign key – which is labeled as sensor_id and id in the two tables, they share the same data type (which is a string with a length of thirty). The foreign key is important because it restricts the storage of any measured data into the database, if the corresponding sensor type is nonexistent in the sensor table. The model of the database is shown in Figure 10. First of all, it was necessary to design a table to store all the vital data about the different sensors available in the MS, like the type, manufacturer, measured unit, or the valid operating range of it. **Storing** information like the previously mentioned parameters is useful because several errors can be found without any human interactions during measurements. The main table of the database contains all of the previously measured data during experimental tests. The records are identified by the sensor_id which was used to measure the value stored in the actual record, along with the date- time attribute, which consists of the date and the precise time when the measurement was taken. Every row contains only one measured value and the corresponding input parameters. The others remain empty. For example, 6V is measured during an RPM test, then the record will be the following:

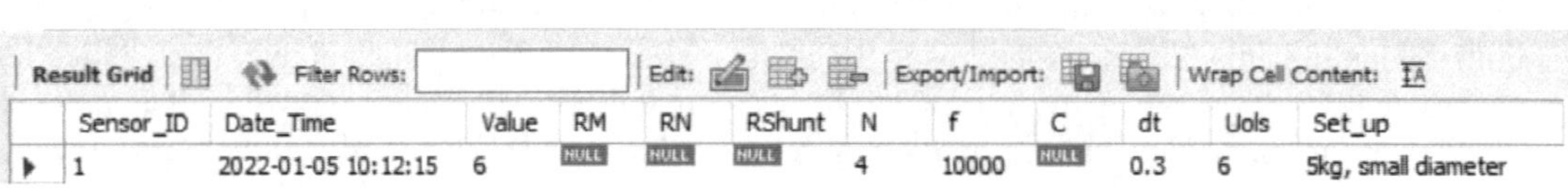

Sensor_ID	Date_Time	Value	RM	RN	RShunt	N	f	C	dt	Uols	Set_up
1	2022-01-05 10:12:15	6	NULL	NULL	NULL	4	10000	NULL	0.3	6	5kg, small diameter

Figure 9. Sample record of the Measured Data table

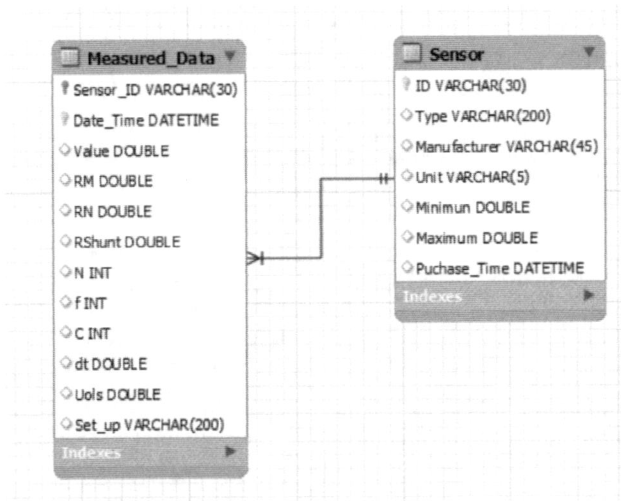

Figure 10. Database model of the underlying MySQL database of the Software Component

4. Summary

The Software Component of an electric motor test bench was realized in the NI LabVIEW environment. The main purposes of the developed Software Component are to separate, collect, store, process, analyze and graphically display the data measured from an electric motor. These data are the angular speed, torque, voltage of the motor, and the intensity of the electric current flowing through it. The NI LabVIEW VI was completed with a Java command-line application for faster pre-processing, a VBA Macro for data analysis, and a MySQL database which serves as a long-term storage place for the measured data. The developed Software Component was partly tested during measurements on a three-phase induction motor, and it has met the requirements while proving to be user-friendly. After having more experience with the Measurement System, we intend to realize additional improvements on the Software Component and complete it with additional features answering the arising demands during the experimental work.

Acknowledgment

„SUPPORTED BY THE ÚNKP-21-3 NEW NATIONAL EXCELLENCE PROGRAM OF THE MINISTRY FOR INNOVATION AND TECHNOLOGY FROM THE SOURCE OF THE NATIONAL RESEARCH, DEVELOPMENT AND INNOVATION FUND."

„SUPPORTED BY THE ÚNKP-21-4-II NEW NATIONAL EXCELLENCE PROGRAM OF THE MINISTRY FOR INNOVATION AND TECHNOLOGY FROM THE SOURCE OF THE NATIONAL RESEARCH, DEVELOPMENT AND INNOVATION FUND."

"THE RESEARCH WAS SUPPORTED BY THE THEMATIC EXCELLENCE PROGRAMME (TKP2020-NKA-04) OF THE MINISTRY FOR INNOVATION AND TECHNOLOGY IN HUNGARY."

References

[1] Szántó, Attila, Sándor Hajdu, and Gusztáv Áron Sziki. "Dynamic simulation of a prototype race car driven by series wound DC motor in Matlab-Simulink." Acta Polytech. Hung 17.4 (2020): 103-22.

[2] Szíki, Gusztáv Áron, Attila Szántó, and Tamás Mankovits. "Dynamic modelling and simulation of a prototype race car in MATLAB/Simulink applying different types of electric motors." International Review of Applied Sciences and Engineering 12.1 (2021): 57-63.

[3] Hadžiselimović, Miralem, et al. "Magnetically nonlinear dynamic model of a series wound DC motor." Przegląd Elektrotechniczny 87.12b (2011): 60-64.

[4] Szántó, Attila, András Szántó, and Gusztáv Áron Sziki. "Review of the modelling methods of series wound DC motors." Műszaki Tudományos Közlemények 13.1 (2020): 166-169.

[5] Sziki, Gusztáv Áron, et al. "Experimental investigation of a series wound dc motor for modeling purpose in electric vehicles and mechatronics systems." Measurement 109 (2017): 111-118.

[6] Szántó, A. et.al: Dynamic Test Measurements and Simulation on a SeriesWound DC Motor. Appl. Sci. 2021, 11, 4542. https://doi.org/10.3390/app11104542

[7] What is LabVIEW? https://www.ni.com/hu-hu/shop/labview.html

[8] Elliott, C., Vijayakumar, V., Zink, W., & Hansen, R. (2007). National instruments LabVIEW: a programming environment for laboratory automation and measurement. JALA: Journal of the Association for Laboratory Automation, 12(1), 17-24.

[9] Schildt, H., & Coward, D. (2014). Java: the complete reference (p. 1312). New York: McGraw-Hill Education.

Development of a measurement method to determine rolling resistance

S Pálinkás[1], Á Tóth[2]

[1,2]University of Debrecen, Faculty Engineering, Department of Mechanical Engineering; 4028 Debrecen, Ótemető u. 2-4., Hungary;
[1]palinkassandor@eng.unideb.hu, [2] totar98@icloud.com

Abstract. Nowadays, car manufacturers tend to produce cars of low consumption to comply with stricter and stricter environmental protection standards. Fuel consumption can only be reduced by decreasing losses emerging during vehicle operation. This is corroborated by the fact that due to the terminal capacity of batteries of electric cars, manufacturers aim to reduce losses to enhance their range of operation. The main goal of our current research is to work out indirect methods of measurement to define rolling resistance, one of the most decisive losses. Later on, the values defined will serve as input data to model vehicle energetics calculations. During our experiments, measurements were carried out with both summer and winter tires. During the evaluation losses of the engine, rolling resistance and air resistance were all considered, while losses from the driving components of the vehicle were ignored.

1. Introduction

Rolling resistance is basically caused by the motion of tires. The running surface is deformed and distorted with every rotation of the tire. Due to being in contact with the road surface during a change in shape, the tires warm up and that mechanical energy turns into heat. Most of the rolling resistance of car tires emerge in this process. Part of the energy passed on to the wheels is consumed by the tires in the process, which leaves less energy for advancing the vehicle. Thereby, the greater the rolling resistance of the tires is, the more it consumes the mechanical energy, provided by the engine. Rolling resistance cannot be eliminated completely, but decreasing it as much as possible means a considerably lower fuel consumption of the vehicle. In other words, reducing rolling resistance plays an important role in reducing fuel consumption.

It is important to note that fuel can only be saved by reducing the losses during operation. These losses are as follows [1]:

- rolling resistance,
- air resistance,
- inertial resistance,
- gradient resistance,
- side force resistance,
- transmission loss,
- losses from the use of auxiliary equipment,
- engine friction.

Factors influencing rolling resistance are the following: the structure of tires, sizes and temperature of tires, air pressure in tires, speed and mass of vehicle, material and pattern of tires, and the slippage of tires. In recent years, several researches [2-4] are focused on determining the above factors. More equations have been defined to calculate rolling resistance. Studies focusing on characteristics of rolling losses of car tires have defined the following equation [5]:

$$f_r = \frac{R_x}{W} = C \cdot \frac{W}{D} \cdot \sqrt{\frac{h_t}{w}} \tag{1}$$

where: R_x= rolling resistance force,
W= weight on the wheel,
C = constant reflecting loss and elastic characteristics of the tire material,
D = outside diameter of tire,
h_t= tire section height,
w = tire section width.

From this formulation, rolling resistance is seen to be load sensitive, increasing linearly with the load. Larger tires reduce rolling resistance, as do low aspect ratios (ht/w). Some confirmation of the general trends from this equation appears in the literature from studies of the rolling resistance of conventional passenger car tires of different sizes under the same load conditions. Other equations for the rolling resistance coefficient for passenger car tires rolling on concrete surfaces have been developed. The variables in these equations are usually inflation pressure, speed, and load. The accuracy of a calculation is naturally limited by the influence of factors that are neglected. At the most elementary level, the rolling resistance coefficient may be estimated as a constant. Table 1. below lists some typical values that might be used in that case [5].

Table 1. Example values for rolling resistances [5]

	Surface		
Vehicle Type	Concrete	Medium Hard	Sand
Passenger cars	0.015	0.08	0.30
Heavy trucks	0.012	0.06	0.25
Tractors	0.02	0.04	0.20

Measuring rolling resistance goes back several centuries and emerges from the interest of the military to decrease horsepower needed to tow cannons [6]. Great scientists such as Coulomb and Reynolds took part in elaboration. In the past 50 years, standard processes have been created to measure rolling resistance. In 2011 Sandberg [7] classified measuring techniques of rolling resistance in the group below:

- Drum tests of tires,
- Trailer methods,
- Coast-down methods,
- Fuel consumption methods.

The drum test is ideal for the testing of tires in the laboratory. Trailer methods add the variation of the road surface to the testing and also monitor some of the transmission loss. Coast-down methods include further properties of the vehicle, while fuel consumption methods measure all the energy losses experienced by the vehicle [8].

One topic in mechanical engineering which integrates vehicle dynamics with technical experimentation is coast-down testing. Coast-down testing is the process of accelerating a vehicle to a high speed on a flat, straight road and coasting in neutral down to a low speed. By recording the amount of time, the vehicle takes to decelerate, it is possible to obtain a model of the loss-inducing forces affecting the vehicle. Obtaining valid coast-down results requires several steps, including experimental planning, data collection, and data processing [9].

2. Elaborating an experimental plan to coast-down testing

2.1. Choosing the right stretch of road for measurements

To test rolling resistance, it is important for the stretch of road chosen to be straight while long enough for the driver to be able to accelerate the vehicle to 80 km/h, so that it can roll on to a full stop after being put into neutral gear. Another important criterion is to choose a less busy road, that is why a road used for agricultural purposes was chosen on the outskirts of Hajdúböszörmény (Figure 1.). The road chosen was not very smooth and was full potholes due to being used by agricultural machines.

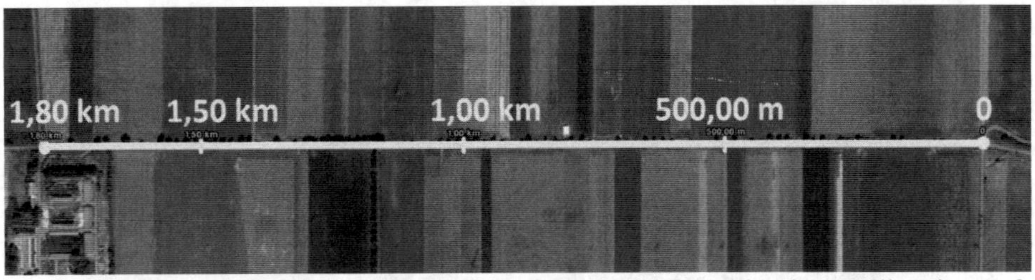

Figure 1. The stretch of road chosen to test coast-down as is shown by the satellite picture from google maps

2.2. Elaborating the aspects of measurement

During the experiments it is highly important for measurement conditions to be identical, whereby the following system of aspects has been elaborated:

- measurements should be on the same day,
- in the same location,
- in the same weather conditions (temperature, humidity, direction of wind, velocity of wind),
- same load on axis,
- same size tires and air pressure,
- closed windows,
- full fuel tank.

2.3. Experiments to test coast-down with summer and winter tires

During our experiments, our main goal was to define rolling resistance with winter and summer tires to obtain a significant difference. During the evaluation, engine losses and losses due to rolling and air resistance were accounted for. The losses emerging from the drive components of the vehicle were ignored because they were considered insignificant. During the experiments, the time needed for a full stoppage and the changes in engine power were measured.

3. Conditions of measurements

3.1. Environmental conditions of measurements

During the measurements, precise weather data is indispensable, which were obtained from the Hungarian Meteorological Service on the day of testing. Data on rain, temperature and relative humidity were obtained from the automatic station operated in Balmazújváros. The data of wind and air pressure were defined based on measurements from the meteorological station in Folyás village. During the measurements, the current temperature, which was 16 °C, was defined utilizing the thermometer installed in the vehicle.

3.2. The measuring device used for the measurements

While defining the exact environmental data, it is indispensable to use a precise method to define the speed of the vehicle while testing coast-down. Nowadays, OBD-II sensor diagnostics are widely used to monitor vehicle operation and consumption [10-11]. An ELM327 Vgate Scan, Advanced OBD2 Bluetooth reading device (Figure 2.) was used for testing, while applying an android-based recording software that could be downloaded free of charge. The module can be fitted into the OBD-II port of a vehicle, which makes it possible to obtain data from the CAN bus of the vehicle. The unit used supports all OBD-II standards: ISO15765-4 (CAN), ISO14230-4 (KWP2000), ISO9141-2. This ensures access to further variables, such as the number of revolutions, power, fuel consumption, operational temperature, etc. in addition to checking sampling frequency. An even more precise procedure to measure vehicle speed precisely would be external sensors, but it would increase both the cost and complexity of the experiment.

Figure 2. The device used for the measurements, ELM327 Vgate Scan Advanced OBD2

3.3. *Specifications of the vehicle used for the measurements*

Table 2. The details of the experimental vehicle

List of specifications make of vehicle Suzuki Liana 1,6; 2005	
Air resistance coefficient	0,5
Perpendicular surface of airflow	2,25 m^2
Mass (own mass of vehicle + 2 persons)	1225 + 105 + 70 kg
Speed aimed at coast-down test	80 km/h
Air pressure and sizes of tires	2,2 bar; 185/65 R14
Type of summer tires	Nordexx Fastmove 3
Type of winter tires	Momo W-1 North Pole

4. The measurement procedure

A specific procedure for coast-down testing can be found in the standard SAE J1263 [12]. The measurement procedure outline below was defined based on the standard and our experimental plan:

1. Prepare vehicle for testing:
    a. Checking vehicle to comply with SAE J1263 standard [12] and production specifications.
    b. Defining test variables: Different seasonal tires were used in our current paper.
    c. Warm-up procedure: The vehicle is driven at 80 km/h for 30 minutes before measurement.

2. Fixing the initial environmental data:
   a. Recording variables: environmental temperature, air pressure, parallel and cross-directional wind speed and maximum wind speed observed.
   b. Should any of the variables above exceed the limit of SAE J1263 [12] the test cannot continue.
3. Testing coast-down:
   a. If the data collecting system is on standby and the staff carrying out experiments is ready for measurement, the data collecting system has to be started to setup operational characteristics of the vehicle. Then the vehicle is accelerated up to 80 km/h along a straight plain smooth road.
   b. After reaching the desired speed the accelerator pedal is let up and the clutch pedal is pressed to put the vehicle into neutral gear. The fall in the number of revolutions while changing into neutral gear is considered while processing data. This is necessary due to the impression of the person's reaction, carrying out measurement and the feedback of devices.
   c. The vehicle must be kept in neutral gear along a straight track while coast-down testing is taking place. Should any dangerous situations arise, such as objects appearing on the road, the test has to be aborted immediately and restarted later.
   d. Once the speed of the vehicle is at 0 km/h, the data collecting system is stopped and the data files are saved.
4. Repeating testing:
   a. Step 3 tests are repeated by using both winter tires and summer tires 3 times.
   b. All together 6 experimental measurements are carried out to get a full picture of the operational characteristics of the vehicle, while using different seasonal tires.
5. Recording environmental data:
   a. After carrying out a series, which consists of 6 experimental measurements, the weather conditions are recorded.
   b. It is checked if the weather conditions are within the limitations of SAE J1263 [12]. If there are data outside of the limitations step 3 has to be repeated until all 3 relevant running procedures are recorded.

5. Measurement results

The measurements were carried out according to the description above. Three measurements were carried out with summer tires, then winter tires were used. The measurements were carried out along the same stretch of road with both summer and winter tires, whereby influencing factors emerging from potential level differences were avoided. Results of the measurements with summer tires are given in Figure 3., whereas those of the winter tires are summarized in Figure 4. In diagrams on the right, the power of the engine and the speed of the vehicle is shown. In the diagram on the left the total fuel consumption and distance travelled are demonstrated. They show that the vehicle covered a shorter distance with winter tires due to the obviously greater rolling resistance.

IOP Conf. Series: Materials Science and Engineering 1237 (2022) 012013 doi:10.1088/1757-899X/1237/1/012013

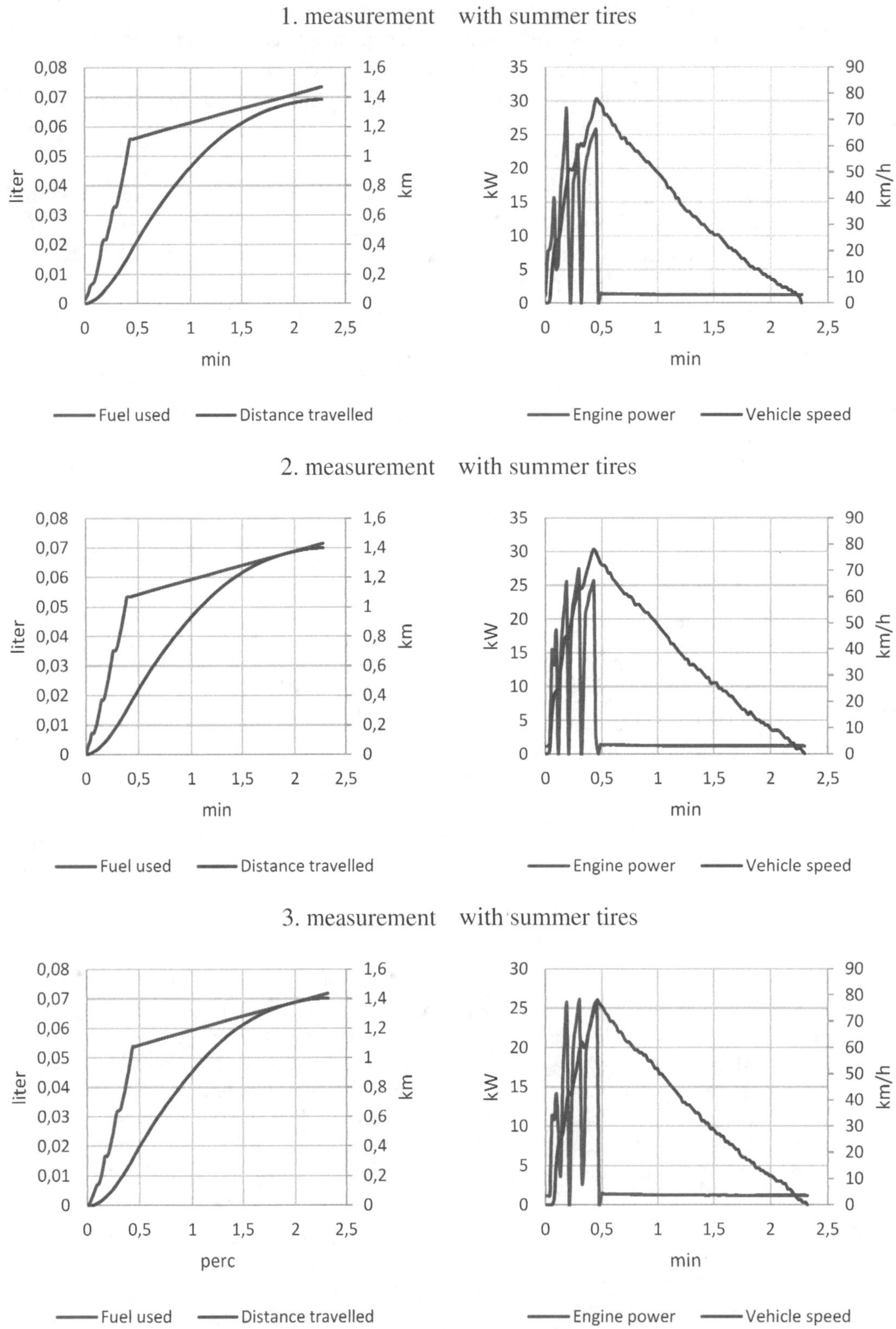

Figure 3. Measurements with summer tires

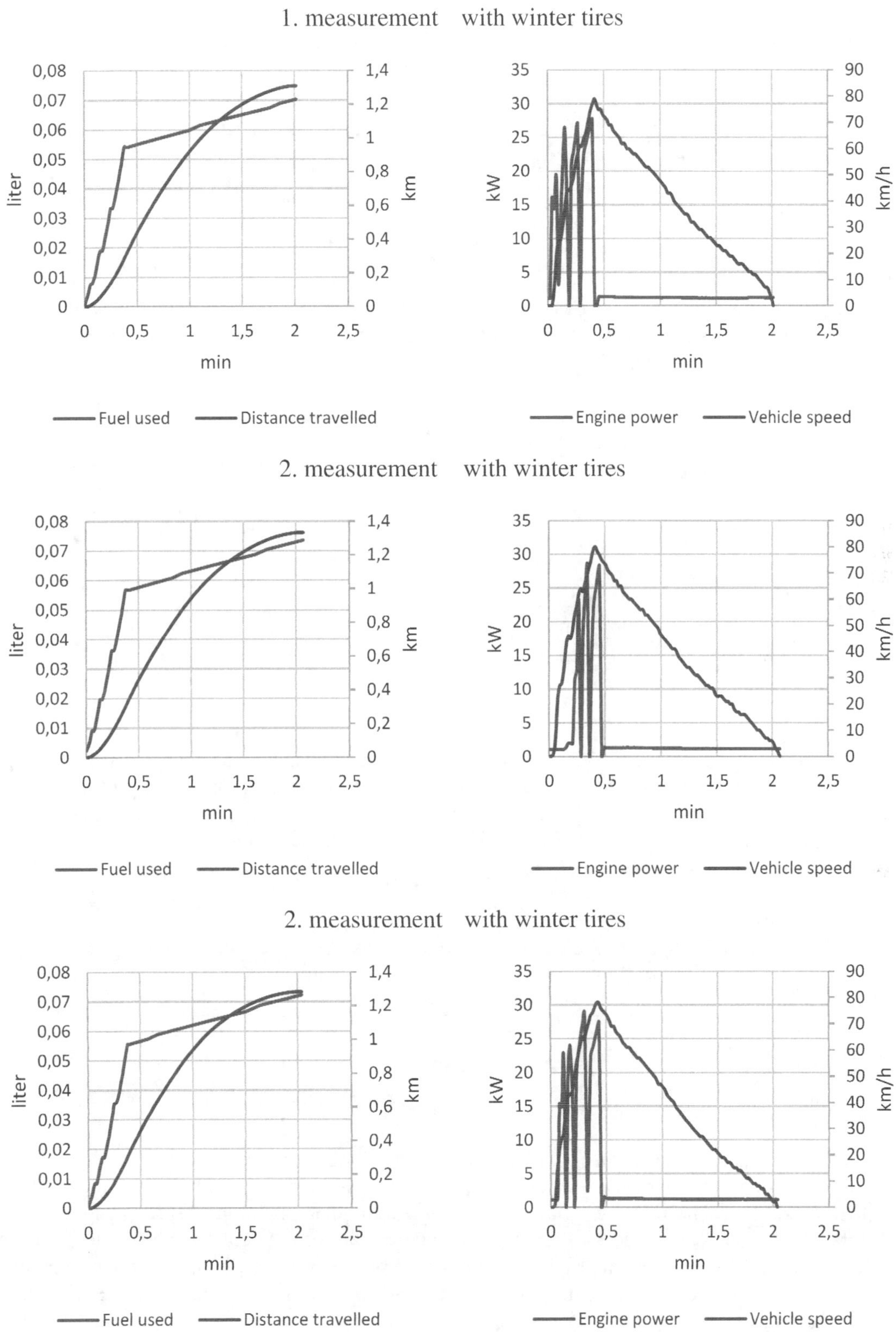

Figure 4. Measurements with winter tires

6. The measurement evaluation

The time needed for complete coast-down was defined in each case based on the measurements carried out as shown in Table 3.

Table 3. Overall time needed for complete coast-down [s]

	measurement 1	measurement 2	measurement 3	average
Summer tires	110,49	113,41	113,12	112,34
Winter tires	96,96	100,51	98,63	98,7

While evaluating the measurements we concluded that we did not manage to achieve the 80 km/h defined experimental plan in each case during the experiments because the driver happened to switch to neutral gear too soon. Based on this fact, we chose 78 km/h as the basis of the calculations as this was the only data available in all the six cases.

During our calculations we defined the air and rolling resistance. While calculating air and rolling resistance the data in Table 2 were considered.

The air resistance force (F_L) and the power (P_L) to cope with air resistance were calculated as follows:

$$F_L = \frac{\rho}{2} \cdot A \cdot c_w \cdot v^2 \quad (2)$$

where: F_L = air resistance force,
ρ = density of air on 15 °C,
A = perpendicular surface of air-flow,
c_w = air resistance coefficient,
v = vehicle speed.

$$P_L = F_L \cdot v \quad (3)$$

It can be observed in Figures 3 and 4 that the power of the engine was always constant when the vehicle was rolling freely, which was considered the loss of the engine. The power of the engine was measured, the air resistance was calculated in compliance with equation 2 and the internal loss of the engine was considered, whereby the output needed to cope with the rolling resistance could be defined. Figure 4 shows the percentage rate of engine power needed to cope with various losses at a speed 78 km/h.

Table 4. The percentage rate of engine power used to cope with various losses

	Rolling resistance [%]			Air resistance [%]			Engine loss [%]		
	1.	2.	3.	1.	2.	3.	1.	2.	3.
Summer tires	67,56	67,33	67,65	27,2	27,4	27,1	5,24	5,27	5,25
Winter tires	71,31	71,86	70,11	24,3	23,9	25,5	4,29	4,24	4,39

The differences in the table can be explained by the fact that the winter and summer tires are different both in terms of their patterns and material compositions allowing them to be applicable in the right weather conditions. The most important property of a summer tire is to ensure safety at a higher temperature on wet road surfaces, whereas the winter tire ensures safety on snowy and slushy road surfaces. Its dense grooves facilitate adhesion on a snowy road surface, which does not change in cold weather either, owing to its composition. This is the reason why most of engine power was used to cope with the rolling resistance during the experiments.

As the power (P_G) needed to cope with rolling resistance at a given speed was defined, the force (F_G) needs of rolling resistance could be calculated as follows:

$$F_G = \frac{P_G}{v} \tag{4}$$

The knowledge of rolling resistance force needs allowed us to define the rolling resistance factor (f) according to the following equation:

$$f = \frac{F_G}{m \cdot g} \tag{5}$$

where: m = the mass of vehicle with the persons carrying out the measurement included,
g = gravitational constant.

Table 5. The value of calculated rolling resistance factor, f

	measurement 1	measurement 2	measurement 3	average
Summer tires	0,058	0,057	0,059	0,058
Winter tires	0,067	0,069	0,065	0,067

The values in Table 5 coincide with the ones given in the literature, which is proved by the fact that the rolling resistance factor value in Table 1 in case of a passenger car and a poor-quality road (medium hard) is 0,08. Only a slight difference can be observed compared to data in the literature, which can be explained by the fact the texture of the stretch of road chosen, where the test was carried out was somewhat different from the road surface given in the special literature.

7. Conclusion

Our main goal defined at the beginning of our research, to provide a measurement method to define rolling resistance has been achieved. Both winter and summer tires were used during our coast-down tests. A significant difference could be shown during the evaluation in terms of the time needed for total coast-down. The engine power percentage rate needed to cope with different losses. The value of the rolling resistance factor between seasonal tires were shown. Although the rolling resistance can be increased by reducing air pressure in the tires, the air pressure was 2,2 bars with both summer and winter tires during the test. This proves clearly that coast-down time is reduced with winter tires owing to the structural formation, material composition and patterns of the wheels. Fuel consumption also appears in the measurement diagrams of the coast-down test, which suggests that the vehicle covered a shorter distance during the test carried out with winter tires. This is another potential method to measure rolling resistance, more specifically the fuel consumption method, which is not specified in detail in the test, but we are going to carry out tests in future to focus on this issue too.

References

[1] Hammarström, Ulf; et al. 2009 *Road surface effects on rolling resistance – coastdown measurements with uncertainty analysis in focus* (ECRPD Deliverable D5a) p 2
[2] Beata Świeczko-Żurek, Piotr Jaskula, Jerzy A. Ejsmont, Agnieszka Kędzierska, Paweł Czajkowski 2015 *Rolling Resistance And Tire/Road Noise On Rubberized Asphalt Pavement In Poland* (Proceedings of the Rubberized Asphalt-Asphalt Rubber Las Vegas, USA)
[3] Wiegand, B., 2016 *Estimation of the Rolling Resistance of Tires* (SAE Technical Paper 2016-01-0445, doi:10.4271/2016-01-0445)
[4] Yudhidya Wicaksana, et al. 2011 *Determining Rolling Resistance Coefficient on Hauling Road Using Dump-truck in Open Pit Coal Mine* (International Symposium on Earth Science and Technology)

[5] Thomas D. Gillespie 1992 *Fundamentals of Vehicle Dynamics* (Society of Automotive Engineers) pp 116-117
[6] Luchini, J. R., and Simonelli, L. A. 1983 *Rolling resistance tests on smooth and textured-surface test wheels* (Rubber division symposia, Vol. 1, American Chemical Society, Rubber Division, Akron, OH) p 140
[7] Sandberg, U., et al. 2011 *Rolling resistance—Basic information and state-of-the-art on measurement methods* (Models for Rolling Resistance in Road Infrastructure Asset Management Systems (MIRIAM)) p 12
[8] Lasse G. Andersen; Jesper K. Larsen; Elsje S. Fraser; Bjarne Schmidt; and Jeppe C. Dyre 2015 *Rolling Resistance Measurement and Model Development* (Journal of Transportation Engineering, 141(2): 04014075, American Society of Civil Engineers) pp 1-2
[9] Cai, Z. T., & Worm, J. J., & Brennan, D. D. 2012 *Experimental Studies in Ground Vehicle Coastdown Testing* (2012 ASEE Annual Conference & Exposition, San Antonio, Texas. 10.18260/1-2—21370) p 2
[10] Dimitrios Rimpas, Andreas Papadakis, Maria Samarakou, 2020 *OBD-II sensor diagnostics for monitoring vehicle operation and consumption* (Energy Reports) Volume 6, Supplement 3, pp 55-63,
[11] Medashe Michael Oluwaseyi , Abolarin Matthew Sunday, 2020, *Specifications and Analysis of Digitized Diagnostics of Automobiles: A Case Study of on Board Diagnostic (OBD II)*, (International Journal of Engineering Research & Technology (IJERT)) Volume 09, Issue 01
[12] SAE 2010 *Road load measurement and dynamometer simulation using coastdown techniques* (SAE J1263, Society of Automotive Engineers)

Acknowledgement
Project no. **TKP2020-NKA-04** has been implemented with the support provided from the National Research, Development and Innovation Fund of Hungary, financed under the **2020-4.1.1-TKP2020** funding scheme.

Design Issues for Slot-Die Coating Heads - Case Study

Ferenc Sarka[1 (0000-0003-3136-4248)], **Zsolt Tóbis**[1 (0000-0002-4711-2224)]

[1] University of Miskolc, Institute of Machine and Product Design, 3515, Hungary, Miskolc.

e-mail: machsf@uni.miskolc.hu

Abstract. In this article, the authors briefly present the development, features, operating principles, and typical applications of slot-die technology. Slot-die technology is a coating method in which a consistent thickness, coating layer, can be applied to a receiving surface in an almost completely closed system. The technology was first used in the production of films but is now used for the application of a wide variety of coatings (e.g., plastics), from battery production, through semiconductor production, to solar cell production. It can also be used to make membrane air filters / filters, which can help to overcome today's covid epidemic. Due to the proliferation of electric cars, the production of batteries has become an increasing slice of industrial production. In this article we present in detail, in a systematic way, the information gathered during the literature search, which we used in our work. In this article we present with a case study, the problems and experiences that arose during the design and manufacture of a coating machine operating under laboratory conditions. The construction of the machine was done with the technical assistance of a joint project with an industrial partner. The laboratory machine will be suitable for sheet-to-sheet coating. The article presents strength and deformation simulations along with the experience gained during the production of the coating machine.

1. Introduction

The invention of slot-die coating technology is attributed to Albert E. Beguin, who filed his patent with the U.S. Patent Office in 1952 (US2681294A). The technology first appeared at Eastman Kodak Co. in the United States [1] as one of the technologies in film production. Nowadays, with the proliferation of electric cars, the production of high-performance batteries is becoming increasingly important. In this area too, it is important to be able to create thin layers on the surfaces of battery electrodes. Slot-die technology is an excellent way to solve this problem. More and more new investments are starting in Hungary, which are related to battery production in some way; so researching, understanding and getting to know the technology can be an advantage for our country. The basis of slot-die technology is to deliver the material, to be applied to the receiving surface (substrate), through a thin gap. The layer thickness of the thin coating can range from a few tenths of a micrometer to a few tenths of millimeter. The liquid flowing out of the gap reaches the substrate as a curtain. There is relative movement between the substrate and the gap. The movement is usually performed by the substrate, compared to where the gap is located. The substrate is very close to the gap. In the original invention, Beguin created his movement using a rotating drum. This type of technique is called a roll-to-roll technique, abbreviated R2R. If the receiving surface is a flat sheet, it is called sheet-to-sheet technique, shortly the S2S technique. It is now possible for the gap to move over the receiving surface. With slot-die technology, both types have their own applications.

Our aim is to create a coating machine with slot-die technology to support our joint project with our industrial partner. By acquiring the initial steps on a small machine, we want to produce machines for industrial production.

2. The principle of slot-die technology

Nowadays, slot-die technology is becoming more widespread, both in industrial and laboratory applications. Figure 1 shows a schematic of a slot-die tool. The names of the tool components and the parts of the coating bead formed at the application site, can also be identified from the figure.

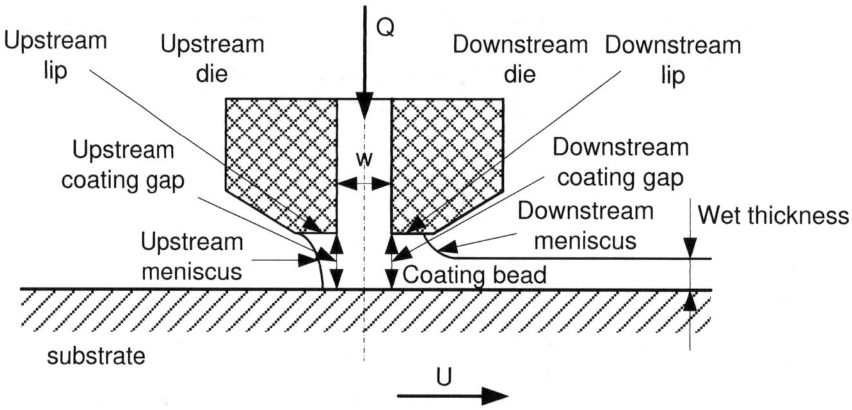

Figure 1. The overall design of a slot-die head

- Substrate: a carrier/receiver surface, to which the coating is applied. It can be rigid or flexible.
- U: the velocity of the substrate.
- Upstream die: the upstream part of the slot-die head.
- Downstream die: the downstream part of the slot-die head.
- Upstream meniscus: the fluid profile of the upstream side.
- Downstream meniscus: the fluid profile of the downstream side.
- Coating bead: a layer of fluid applied from the slot-die head.
- Wet thickness: the thickness of the applied material when wet.
- w: the width of the gap.

The fluid flowing through the gap arrives at the receiving surface to form a consistent thickness. Changing the size of the thickness can be influenced by the values of several application parameters. Such parameters are:

- Density and viscosity of the liquid.
- The speed of the substrate.
- Flow rate of fluid flowing through the gap.
- The size of the gap.

In the case of a given combination of listed parameters, a stable liquid profile is formed on both sides of the slot-die head; in which case it is possible to apply a thin and consistent layer thickness. In his invention, Beguin [1] proposed the use of a vacuum in front of the upstream die, of the slot-die head. The vacuum helps to stabilize the fluid profile exiting the gap, by removing air adhering to the substrate surface as it moves. Applying a vacuum is only a problem during high-speed coating applications, also it is not worth applying below 2 m/s [2,3,4]. In 1976, Ruschak researched the stability of coating production and introduced the concept of the "coating window." If the parameters are inside the "window," the coating will be good [5]. Changes in coating preparation parameters have a major impact on the shape of the upstream and downstream meniscus. The size of the gap between the inlet and outlet sides of the slot-die head is created by a spacer (shim) plate. By changing the thickness of the spacer plate, it is possible to apply different thicknesses (we can influence the thickness with other parameters, but this is the main option). The following figure (Figure 2) shows a disassembled slot-die head, which also shows the names of the inner surface of the head and the fluid distribution channels.

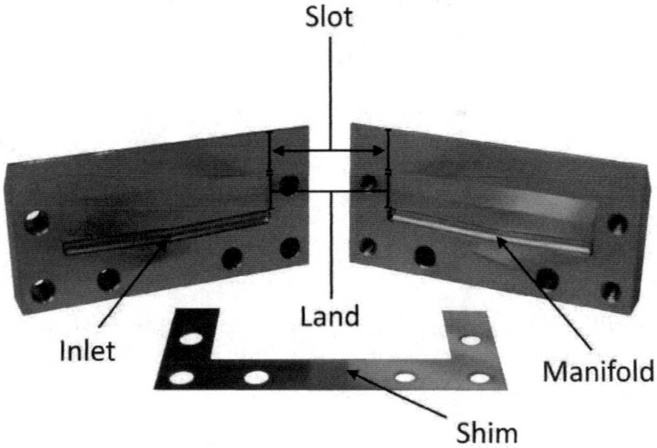

Figure 2. Names and design of the interior of the slot-die head (symmetrical) [6]

3. Slot-die head interior design

If we look at the products of the companies that make slot-die heads, we find that two different principles/directions have emerged in the design of the interior of the head. One direction is when the two halves of the head (upstream and downstream) have the same design (symmetrical head). The other direction is when the downstream side element of the head is a flat surface, while a chamber(s) is formed on the upstream side (asymmetric head). The symmetry or asymmetry of the lips of the head is another very complex issue, which are not discussed in this publication; but there are studies examining the effect of the shape and position of the lips to the upstream and downstream meniscus [3].

3.1. Some characteristics of asymmetric heads

Like all engineering solutions, the asymmetric-symmetrical design of slot-die heads have disadvantages and advantages.

Disadvantages:

- For such heads, two different parts have to be manufactured, which complicates production, but this effect is not significant.
- As the manifold channel passes through the upstream die element of the head, the assembled head must be closed on two sides with end caps. New parts are needed for the design.
- Usually used for wide width heads.
- The direction of the dividing plane of the head should be as horizontal as possible during installation.
- The maximum deformation of the heads due to the pressure in the interior is greater than in the case of symmetrical heads (see 4.3 for details).

Advantages:

- In asymmetrical heads, the manifold channel runs along the entire head to help clean the head.
- It is also easier to set the parameters of the machined surface in a case where the surface to be machined, can be accessible from two directions.
- Machining the inner surface of a non-ducted element is simple, a planar grinding task.
- When filling the head with liquid, any air bubbles that may form can be more easily removed from the head (horizontal placement).

Figure 3. An example of an asymmetrical head design [6]

3.2. Symmetrical heads

An example of a symmetrically shaped head is shown in Figure 2.

Disadvantages:

- Due to the symmetrical design of the head, it is difficult to remove air bubbles from the liquid. Often some additional elements are needed (aeration valve, adjustable head angle).
- Harder to clean because the head is impassable.

Advantages:

- Two parts of the head is identical, must be made of the same parts.
- The manifold channel does not run through the head, so no end caps are required.
- The maximum deformation of the heads due to the pressure created in the interior is smaller than in the case of asymmetric heads (6.3).

When deciding on the design of the heads, you can choose based on the properties indicated above.

4. Case Study

After the introduction, we would like to present a case study in the publication. In 2019, our institute joined a tender project. In the project, we were given the task of designing the mechanics of a non-traditional printing machine. In addition to mechanical engineers, chemists and physicists took part in the project; it was their job to experiment with the special material. At the beginning of the work, we first got acquainted with the coating technologies, from which many types have evolved (Dr. Blade, Microgravure, Slot-Die, spray coating, etc.).

4.1. Starting the planning task

The choice of application technology should always be based on the properties of the material to be applied to the substrate. In our case, the base material was PVA (polyvinyl alcohol) and PU (polyurethane), modified by chemist and physicist colleagues with various ingredients. One of the ingredients is a solid for which particle size and hardness are very important. Large grains can block fluid flow, and hard grains wear the inner surface of the slot-die head; causing the gap to change in size and the application to be uneven. Fortunately, in our case, the wet thickness of the film and the size of the solid grain should not cause this type of effect. It is important to know the solvent of the liquid, as it can be very aggressive and even explosive. Unfortunately, in our case we have this problem, as the solvent for PVA is ethanol and the solvent for PU is toluene; both of which form an explosive mixture with air (see Material Safety Data Sheet). If an explosive atmosphere can develop during the operation of a machine, in Hungary; it is regulated by the National Directorate General for Disaster Management, Ministry of the Interior (NDGDM) [9]. Furthermore, to know the drying kinetics of the materials is very advantageous, because than we really have a good choice of technology. In our case, dealing with the materials' drying time, depends greatly on the amount of solvent, air pressure and temperature.

IOP Conf. Series: Materials Science and Engineering 1237 (2022) 012014 doi:10.1088/1757-899X/1237/1/012014

4.2. Spraying experiment to select coating technology

To obtain first-hand knowledge of the material to be applied and to select the appropriate application method, a spraying experiment was performed. In the experiment, PVA was applied to a PC (polycarbonate) film. Ethanol was used as the solvent. PVA was diluted for spraying (close to water density). An air-brush gun (Fine Art Fa-1800 type with 0.5mm nozzle) was used for spraying. Spraying was performed in an extraction chamber. (Figure 4). During spraying, we found that it was possible to create a coating.

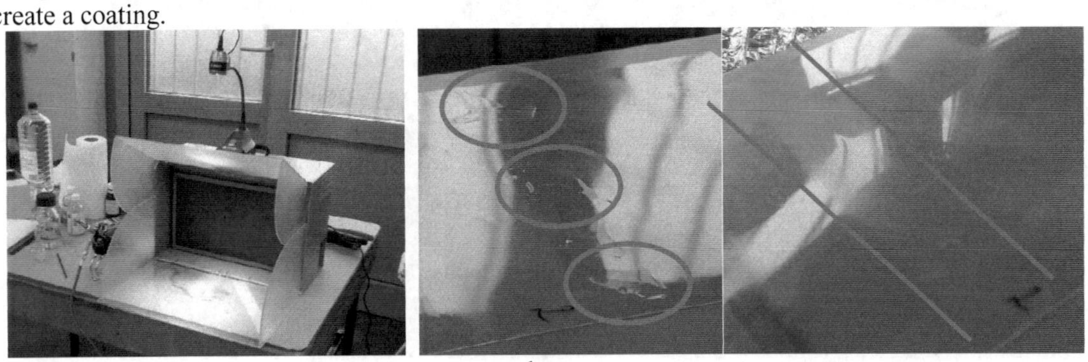

a, b, c,

Figure 4. The spraying experiment a) Arrangement, b) the material did not adhere to the PC foil, c) good coat

Features:

- 15μm thin layer can be created.
- The coating material should be warmed to at least room temperature (raw material should be stored at 5 °C if the material does not adhere to the PC surface when cold [Figure 4. part b]).
- Atomized PVA particles will appear in the air, which is another danger in addition to the solvent.
- The applied coating dries in a few minutes.

Unfortunately, we cannot provide more detailed information because the photo technology material is a secret of our industrial partner. Based on the experiment, we concluded that a closed (or nearly closed) system is required for application. This eliminates the problem of rapid drying and atomized airborne particles. This is how we got to the selection of slot-die technology, which seems to form an almost closed system, according to Figure 1. The experiment highlighted the importance of adhesion between the substrate surface and the applied material. In the absence of mutual "attraction" between the two surfaces, coating may become impossible (the applied material shrinks to droplets, Figure 4, part b marked with circles). The adhesion of the materials can be influenced by treating of the substrate. A detailed description of the applicability and mode of treatment can be found in the literature [8].

4.3. Difficulties encountered in the design and manufacture of laboratory machinery

Once the application technology has been selected, the design and selection of the elements can follow. Further research was limited to slot-die technology. In order to gain experience with slot-die heads, we set out to create a prototype. Here we took advantage of the benefits and disadvantages presented earlier. During the design, we examined the proficiency of the slod-die head elements using FEM simulations. The first version of the slot-die head was a symmetrically designed head. The FEM test result for this is shown in the following figure (Figure 5). The load was the pressure inside the interior, which was chosen to be 1 MPa, which is a small value for the slot-die heads. The fastening was made through the mounting holes. Fixation and constraints are presented on the next figure (figure 5.)

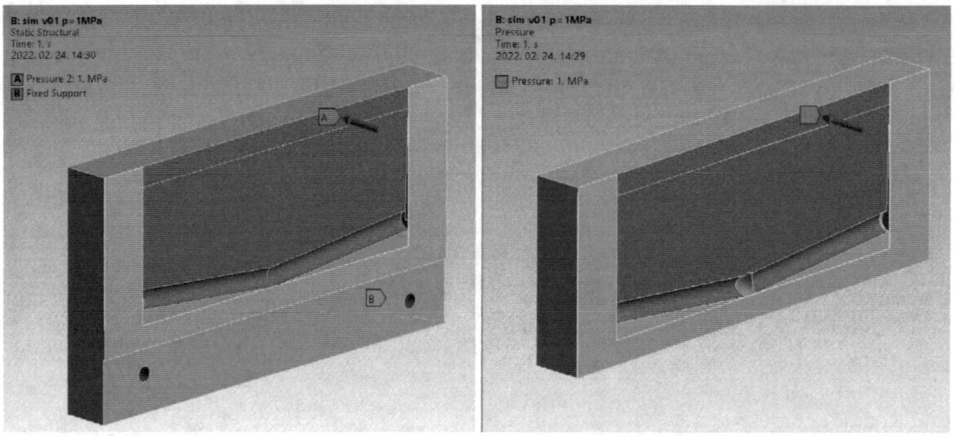

Figure 5. Fixation and constraints

The result of the test was that the strength compliance was met, the maximum stress value was 1 MPa in the tested components (in this case, the yield strength of the aluminum alloy is 280 MPa). However, if we look at the displacement values and the direction of the displacements, we see that the situation is no longer perfect. The pressure on the inner surfaces deformed the outer part of the head. The most problematic place is the environment around the lips (circled in figure 6.). Here you can see that the outer edges of the lips turn outwards, even the inner edges turn inward. This changes the size of the width of the gap in the slot-die head and the shape of the lips.

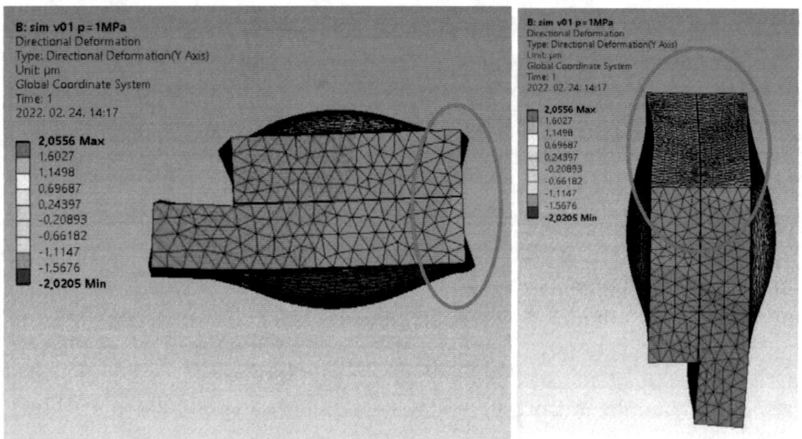

Figure 6. The FEM examination of a symmetrical head

This deformation may already affect the application. The value of the deformation is around 2 µm. And by increasing the internal pressure, this situation will only get worse. Like all results, this should be evaluated. If the gap of the tool is 200-300 µm, the displacement of 2 µm is not significant. On the other hand, if the size of the gap is only 20 µm, then a displacement of 2 µm will result in a deviation of 10 %. Optionally, part of the gap may close, or a solid component of the liquid can stick between the lips. It is recommended to choose stainless steel as the material for the slot-die head. Corrosion and consequent material loss and sizing do not have to be considered. When choosing a material, a material with high tensile strength is recommended to be able to withstand the pressure in the head (more than 1 MPa). Manufacturers recommend 15-5PH stainless steel, in H1025 heat treatment condition, then UTS above 1000 MPa. Unfortunately, high tensile strength is not a guarantee of small deformation, as it is most affected by the modulus of elasticity. This value is lower for stainless steel (1.9x105 MPa) compared to carbon steel (2.1x105 MPa) [11]. The deformation of the lips can be improved by changing the shape of the geometry. The outer edge of the lip, which is turned inwards. If removed, this problem is reduced (see modified head (Figure 8 and Figure 9)). The deformation of the lips affects the change in the size of the gap, which affects the thickness of the applied layer.

4.4. Comparison of symmetrical and asymmetric heads using FEM

After the first test, two types were made with the same dimensions, one with a symmetrical internal design and the other with an asymmetric design.

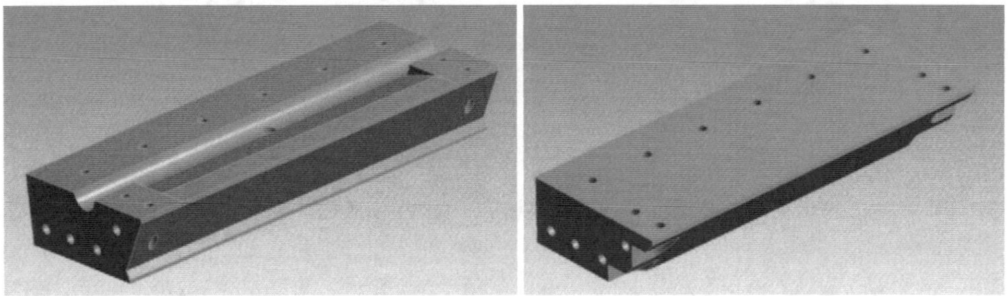

a) b)

Figure 7. Internal design of the two parts of the head (a: part #1, b:part #2)

In both, the symmetrical and asymmetrical case, deformations due to the internal pressure (applied pressure 1 MPa) were examined. In the study, we compared two different supporting states, to look for the effect of different installation options. On the next figure (Figure 8.), the fixations and constraints of the symmetrical head configuration in two different fixation cases can be seen (back, back and bottom).

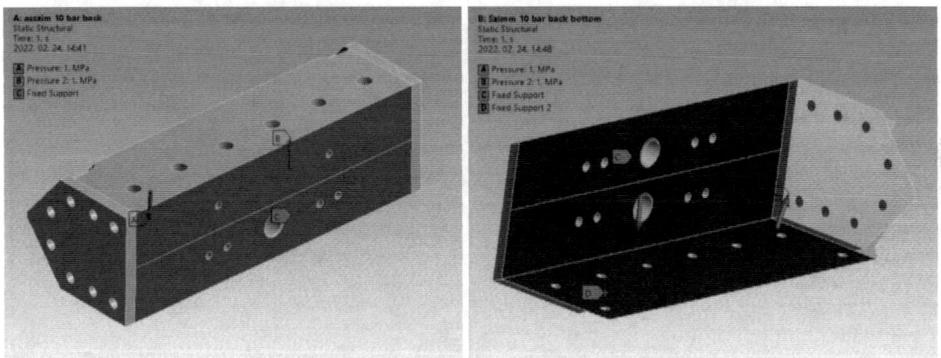

Figure 8. Fixation and constraints

Figure 7. shows the two parts of the head from which the symmetrical and asymmetric versions can be assembled. In part a) the element with the manifold channel; in part b) the element with a flat surface. If we assemble two #1 parts we get a symmetrical head. If we assemble one from part #1 and one from part #2, we get an asymmetric head.

The following two figures (Figure 9 a) and b)) show the results of heads with a symmetrical interior with two different gripping states.

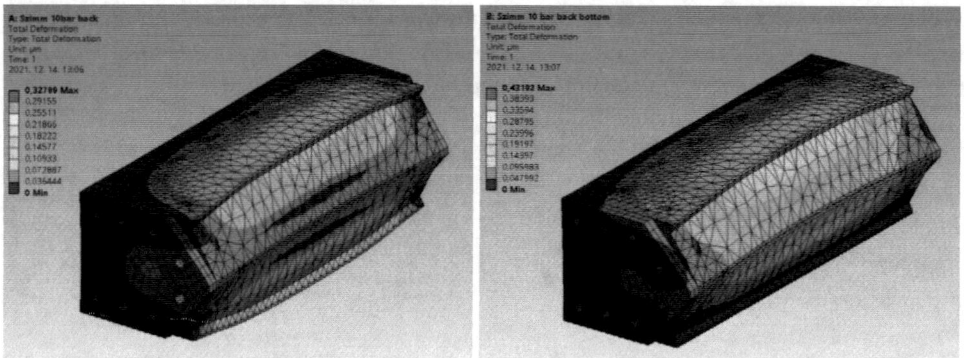

a) b)

Figure 9. FEM test of the symmetrical head, a) supported to the rear surface, b) supported to the rear and lower surfaces

The maximum value of the displacement increases by 0.1 μm in case b) compared to case a). If we look at the displacement values that appear at the lips, we also see an increase there. That is, in the case of symmetrical heads, mounting on a rear surface favors better deformation conditions. The following two figures (figure 10.) show the VEM test result for an asymmetric head with two different gripping cases.

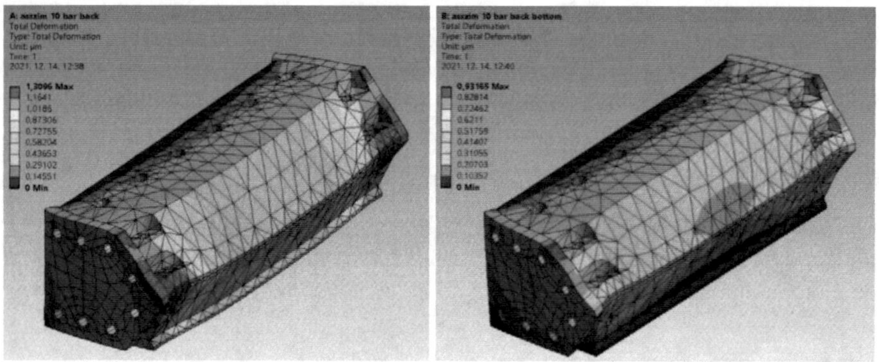

a) b)

Figure 10. FEM test of the asymmetric head (a) affixed to the rear surface, (b) affixed to the rear and lower surfaces

In the case of the asymmetrical head, we find a larger maximum displacement than in the case of the symmetrical heads, but it is concentrated only in the end caps and their environment. The deformation of the lips is the same order of magnitude. If we look at the test with the rear and lower fixation (Figure 9. b), we can see that the deformation of the lips is 0.1-0.2 μm, which is a very similar value to that of the symmetrical heads. Considering that there are several advantageous features in the operation of asymmetric heads, it is worth choosing them to use a strong rear and lower grip. An example of such a design is shown in Figure 11. There is still much room for the optimization of the grips and geometry to achieve significant material/mass savings; an example of which is given in [13]. A generic design can also be included in this area, as we see a similar example of material savings in generative design, using [14], [15], [16] literatures.

4.5. Surface roughness and accuracy, tolerances

The quality of the machining of the slot-die head is also a very important issue. Manufacturers of slot-die heads have developed special grinding processes [10] that can achieve up to Ra 0.02 roughness. It is recommended to have as little surface roughness as possible, so that the bumpiness of the tool lip does not copy onto the applied layer. The flatness of the surfaces is also very important, as the velocity of the fluid flowing through the gap of the tool will only be constant at a constant volume flow. If the gap size changes, it will flow faster in the narrower parts and slower in the wider parts due to the incompressibility of the fluid. This affects the thickness of the applied layer. The more continuous and uniform the flow conditions are, the more uniform the layer is. Manufacturers target a deviation of 2-3 μm for the lips of the tool, it is recommended to approach this. After applying the liquid, the surface tension of the liquids can still smooth the surface, but in the case of a fast-drying material, there may not be enough time for it. The surface quality data described above are very strict and expensive to produce. The lips of the heads have a small surface area, and the components of the heads have a significant weight, so the lips must be protected from mechanical impact. It is advisable to use some protective/covering caps that will only be removed from the lips once the heads are in place and ready to begin application.

4.6. Assembling the two halves of the slot-die head

There are several things to look for when assembling the two main components of slot-die heads:

- the order in which the screws are tightened,
- tightening torque,
- to position the elements of the head relative to each other.

The tightening sequence should be as follows; by starting to tighten the clamping screws in the middle of the head, then moving alternately toward the edge of the slot-die head, and then moving alternately toward the lips of the slot-die head. During tightening, check the size of the gap with a feeler gauge. In our case we must realize a 250 μm gap width. For screws, it is advisable to use stainless-steel with a friction-reducing coating (e.g., Teflon). Using this method, a larger proportion of the torque applied to the screw head is used as compression, while a smaller proportion is used to overcome the friction between the running surfaces. Of course, the screw connection must be self-locking, it is worth protecting the connection against loosening. In the case of screw connections, with the repetition of the tightening-loosening cycle, the compressive force due to the tightening torque decreases with each repetition, as can be seen in [12]. We can counteract this phenomenon by stabilizing the frictional conditions. Disassembly and assembly of the heads is a common operation on coating machines, as the head must be cleaned after each shutdown or change to another material, for which disassembly is essential. Optimal tightening torque is between 12-14 Nm. Thus, the clamping between the surfaces is sufficient to seal the surfaces outside of the gap and the shim elements are not overloaded. The elements of the head must be positioned relative to each other. Therefore, orienting elements must be developed.

Figure 11. CAD model of the slot-die head examined

4.7. Design of the machine to be made

The design of the laboratory machine is basically based on BOSCH profiles. Construction from profiles is a common technology for experimental or custom-built machines and equipment. When designing the frame structure of the machine, the goal is to keep deformations at a very low level. Any deformation in the frame of the machine has a detrimental effect on the quality of the application, so heavy-duty sections (90x45) must be used. Basically, we do not encounter large forces during the operation of the machine, in practice we have to take into account most of the loads resulting from the own weight of the elements. The designed machine is a(n) S2S system, and the substrate moves under the slot-die head. The head is mounted on a bridging (bridge) component in a naturally adjustable manner. The table containing the substrate is connected to the frame of the machine by two parallel linear guides and 2-2 carriages per guides. In the case of overlapping elements, it must be possible to adjust (level) so that the table always passes at the same height under the head. The table connects with one fixed and three height-adjustable elements (screw thread). This ensures that the table can be leveled in two directions. The level is adjusted relative to the bridge of the slot-die head, using a hundredths of a millimeter dial gauge. Gauges should be placed so that they measure at both edges of the table. After setting the table, the slot-die head must be set relative to the table. This can be done in two ways. In one case, gaps are placed between the lips of the head and the table (e.g., with gaps of guaranteed thickness manufactured by SKF [17]). Another option is to use two gauges attached to the two edges of the slot-die head. In this case however, the accuracy of their installation is very important. Examples of both solutions are given in the literature. As long as the thickness of the layer to be applied does not need to be changed, the gap can be fixed after an adjustment. The following figure (Figure 10) shows the current state of the experimental machine.

Figure 12. The S2S coating machine in a half-assembled state

5. Summary and future plans

The slot-die technology presented in the above chapters is a forward-looking coating technology. The technology seems very simple at first glance, but it can also be seen from the diversity of the literature used, that it is not at all. In the article, the authors presented the experience gained during their work in designing a slot-die head and a machine that works with it. They presented their experience in the design and construction of the experimental machine. Using simulations, the differences between symmetric and asymmetric slot-die heads were pointed out. The authors continue the work with the design and construction of the fluid delivery system, which is essential for the preparation of the coating. The results found there are summarized in a subsequent article.

6. References

[1] Beguin A E 1952 Method of coating strip material *US patent* No US2681294A.
[2] Chang H M, Chang Y R, Lin C F and Liu T J 2007 Comparison of Vertical and Horizontal Slot Die Coatings *Polymer Engineering and Science* **47** 11 pp 1927-36
[3] Ding X, Liu J and Harris T A L 2016 A Review of the Operating Limits in Slot Die Coating Process *Transport Phenomena And Fluid Mechanics* **62** 7 pp 2508-24

[4] Tanwar J, Vinjamur M and Scriven L A 2007 Design Principles of Integrated Vacuum Slot Arrangement *Wiley Interscience* **53** 3 pp 572-78
[5] Ruschak K J 1976 Limiting Flow in Pre-Metered Coating Device *Chemical Engineering Science* **31** pp 1057-60.
[6] https://www.nordson.com/en/divisions/polymer-processing-systems/products/fluid-coating-systems/ultracoat-adjustable-lip-slot-dies. (last opened: 31.12.2021.)
[7] https://www.ossila.com/pages/slot-die-coating-theory#defects (last opened: 31.12.2021.)
[8] Gilliam M 2015 *Handbook of Manufacturing Engineering and Technology* Polymer surface treatment and coating technologies, ISBN 978-144714670-4 pp 99-124.
[9] Fire Protection Technical Guidline – Explosion Protection, TvMI 13.2:2021.01.15., Ministry of Interior, Emergency Management.
[10] https://www.yasuiseiki.com/slot-die.php. (last opened: 31.12.2021.)
[11] http://matweb.com (last opened: 31.12.2021.)
[12] Talal A S and Lovas L 2021 Bolt Preload Variation in Cyclic Tightening *GÉP* **LXXII** 3-4 pp 9-12
[13] Kamondi L, Szabó F J, Tujner I 2003 Strain constraints in thin for electronic panels *Workshop on optimal design of materials and structures* pp 1-12
[14] Sereg B L, Ficzere P 2021 Weight Reduction and Range Increase of UAV Using Generative Design *Gép* **LXXII** 3-4 pp 57-60
[15] Szabó K 2021 Steps of Generative Design in Integrated CAD Systems *Design of Machines and Structures* **11** 1 pp 53–58
[16] Trautmann L 2021 Product customization and generative design *Multidiszciplináris Tudományok*, **11** 4 pp 87–95
[17] SKF 2018 *SKF Karbantartási és kenéstechnikai termékek*, PUB MP/P1 03000 HU 2018. Augustus.

Acknowledgments
The authors would like to thank our colleague Attila Potyka for assembling the experimental equipment and for his helpful participation in the design process.

IOP Conf. Series: Materials Science and Engineering 1237 (2022) 012015 doi:10.1088/1757-899X/1237/1/012015

Vehicle Dynamic Development and System Analysis Project

Tamás Szakács and András B. Frigyik
Óbuda University
Bánki Donát Faculty of Mechanical and Safety engineering
Institute of mechatronics and vehicle engineering

E-mail: szakacs.tamas@bgk.uni-obuda.hu,frigyik.andras@bgk.uni-obuda.hu

Abstract. The Vehicle Construction Laboratory of the Óbuda University has been funded to develop Shell EcoMaraton, Bosch GoKart, Bosch Electromobile, Emerson Pneubobile, and Formula Student Hungary project cars. Some international cooperation, based on the European Commission's Erasmus+ program, has already taken place in the lab. In the year 2021 a mixed team of Hungarian, and international students (participants of the Stipendium Hungaricum scholarship program) started to work on a project related to Vehicle Dynamic. The scope of the project is to develop a test car, that can be driven in a conventional way or capable of autonomous operation. The short range goals are the following: to build a car, with electric driving and steering capabilities, and to work out the system scheme of sensor, controller, and actuator network, for system development, mathematical analysis, and modelling/simulation.

1. Introduction

The goal of this paper is to present the initial of the student project. Internationalization of Emerson Pneumobil competition has started already in the second year of the competition series, and grew quickly. In the recent years approximately 50 teams are signing up from 9, mostly EU countries. There has been effort taken to involve further-EU (Finland, Germany), and non-EU countries as well, like Serbia, Montenegro, South-Korea and Turkey.

In 2021 a group of team was formed from English mechatronic students mixed from Hungarian, and Stipendium Hungaricum students within the frame of English Vehicle dynamic subject. This project involves further students from non-EU countries. At this point students are developing a concept car to develop electric vehicle protrusion, and vehicle stability control systems, but later in year 2022 they are planned to work on Pneumobile project too.

Until less than two decades ago, the primary focus was on using electric motors in Formula E cars only. Today these engines are used on various types of vehicles. These engines are essential for autonomous vehicles to carry out automated tasks due to the ease of control over them compared to internal combustion engines.

These engines provide the kart with a driving experience similar to Formula One, allowing young individuals interested in this F1 racing field the possibility of a similar platform. Therefore, when replacing internal combustion engines with electric motors, care must be taken to provide similar performance to the previous vehicle and mainly safety.

Charles Jeantud electric car offered excellent performance in the Paris-Bordeaux-Paris race, but the 950 kg batteries did not support delivering a driving experience for a reasonable time. The main barrier of using electric actuators previously was the energy-storing method. This problem prevented this technology from seeing the light despite the actual improvement for both batteries and engines volume over the years. Despite the tremendous technological development that has allowed this technology to be widely used, many drivers still prefer the classic driving approaches. In December 2020, the American research department and advocacy division published consumer interest and knowledge of electric vehicles survey results. This survey showed that only 30 per cent of the participants have knowledge about electric vehicles. However, many professional drivers have shown interest in using electric cars, such as Nelson Piquet Jr, who said that "If you care about the fame and flashes the F1 is the best racing series, but equally exciting for racing drivers are competing in Formula E. Formula E was created to boost electric motorsport, but still very few series are purely electric."

In Section 2 the concept will be introduced, in Section 3 the modell structure, and modell details will be described and Section 4 shows the modelling of the vehicle body. In the conclusions a brief summary, and outlook can be found.

2. The concept of electric go-kart

One of the most important advantages of changing the previous structure is to provide static stability of the vehicle since the internal consumption motor was located to the right of the kart making the turn manoeuvre to be unstable. Moreover, the BLDC motors have a compact size compared to the internal consumption engine and a better power to weight ratio. The developed kart contains a separate motor that will drive each rear wheel, and two front wheels provide static stability and control the steering. Figure 1 also shows the potential position of the battery. The battery is considered relatively heavy. Therefore, it might affect the kart balance when choosing the wrong position. It is also essential to consider safety reasons since the battery might explode when damaged. Due to this reason, the proposed position is least exposed to the other elements and ensures the best safety to the battery.

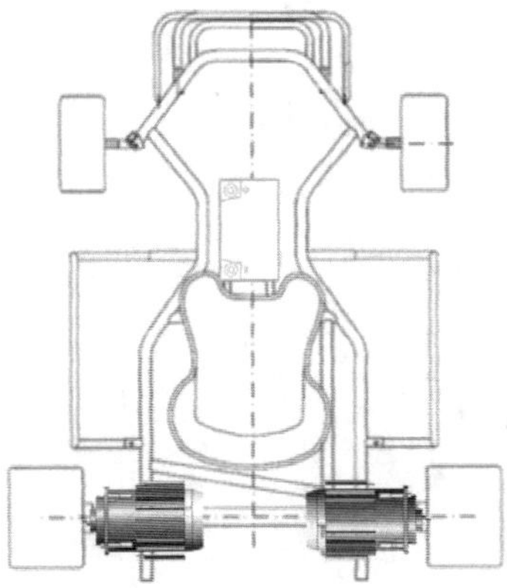

Figure 1. The electric model with the actuator distribution and battery.

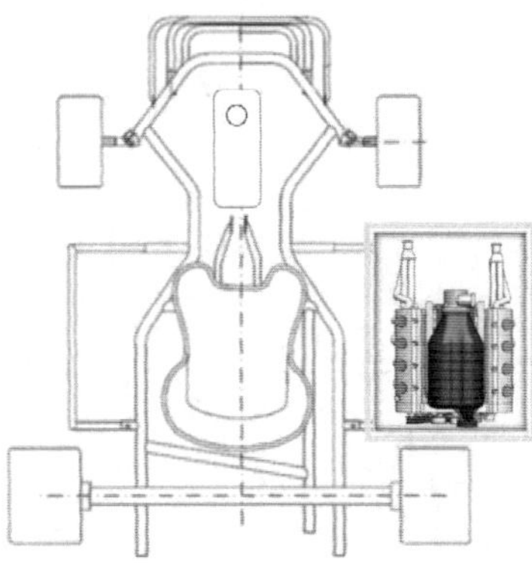

Figure 2. The internal combustion engine kart with gasoline tank.

The used electric motors are brushless DC motors. A BLDC motor works similarly to a DC motor according to the Lorentz force law except that in a BLDC motor, the current-carrying conductor is fixed while the permanent magnet is the rotating part. The permanent magnets can vary from two to eight pole pairs with alternate North (N) and South (S) poles based on the required magnetic field density in the rotor, the proper magnetic material is chosen to make the rotor [1]. In the DC commutator motor, the current polarity is reversed by the commutator and the brushes, but in the brushless DC motor, the polarity reversal is performed by semiconductor switches which are to be switched in synchronization with the rotor position [2]. These motors have a better mechanical structure compared to the conventional DC motors where the permanent magnet rotor provide high efficiency. Due to operating without brushes, the motors will be able to operate at higher load speeds and less friction. In addition to a long working life and lower maintenance requirements.

3. Model structure

Figure 3 shows the top layer of the OMC concept car controller structure.

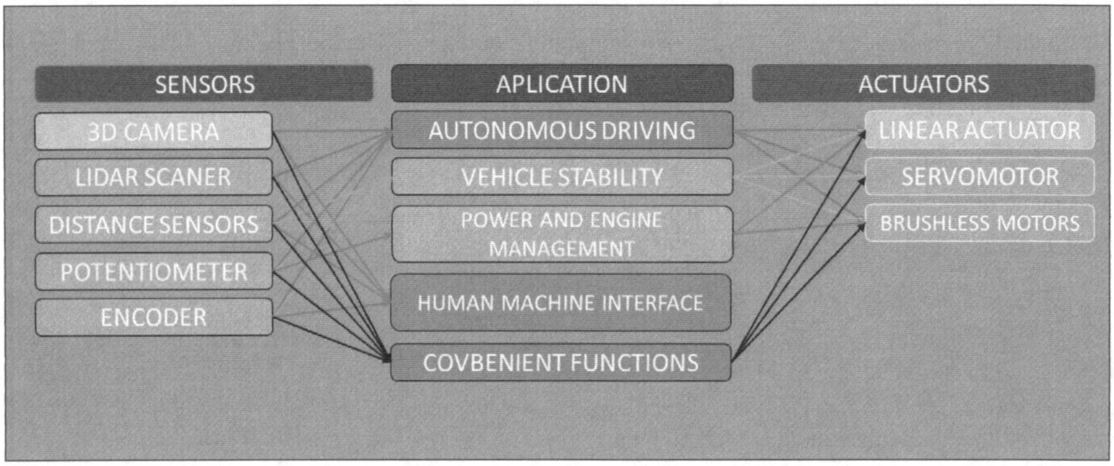

Figure 3. Controller concept of the OMC car.

The system contains sensors, which are providing data for applications, which are influencing actuators.

Table 1. Some features related to the design.

The sensors used:	Applications:	Actuators:
- 3D camera	- Autonomous driving	- Linear actuators
- Lidar scanner	- Vehicle Stability	- Servomotors
- Distance sensor	- Power and Engine Management	- Brushless motors
- Potentiometer	- Human Machine Interface	
- Encoder (angle sensor)	- Convenient Functions	

IOP Conf. Series: Materials Science and Engineering 1237 (2022) 012015 doi:10.1088/1757-899X/1237/1/012015

Functions currently under development:

- Microcontroller (Raspberry, Arduino, Mini-pc)
- Throttle sensor
- Gyroscope
- Steering (angle) sensor
- Braking sensor

3.1. ESP

The electronic stability program (ESP) supports the driver in nearly all critical driving situations (Figure 4). It comprises the functions of the antilock braking system (ABS) and the traction control system, but can do considerably more. It detects vehicle skidding movements, and actively counteracts them. The planned structure of this function is shown on Figure 5.

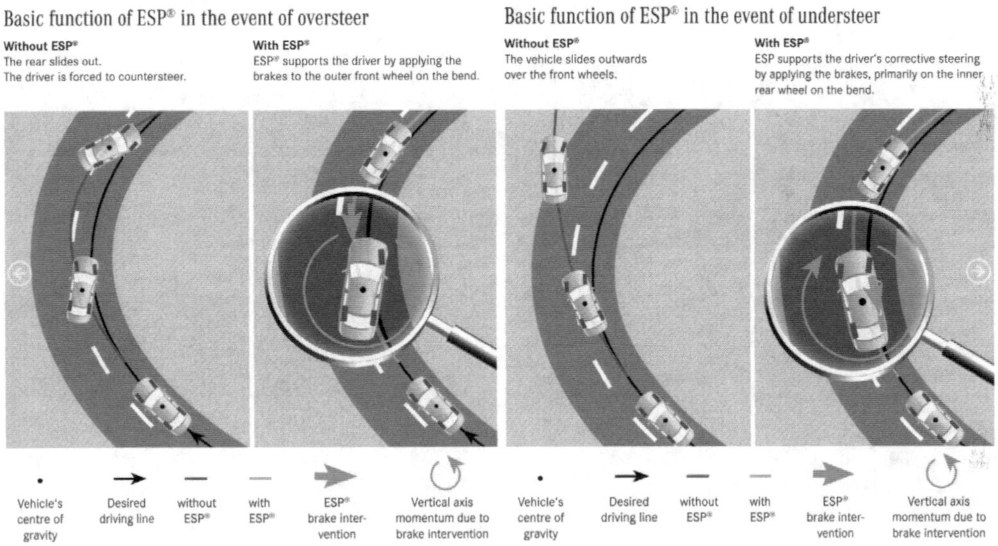

Figure 4. Basic ESP functions.

3.2. Surrounding environment perception and mapping [4],[5]

INVOLVED SENSORS: The mentioned task is going to be included in the main computer on-board the vehicle (OMK) using the facilities provided by the Robot Operative System core (ROS) and the perception sensors inside the Autonomous Vehicle (AV) as describes on Figure 6.

3.3. Anti-drunk driving car safety system

Description of the Alcohol Detection with Vehicle Controlling system: The main purpose of this project is "Drunk driving detection". Nowadays, many accidents are happening because of the alcohol consumption of the driver or the person who is driving the vehicle. Thus Drunk driving is a major reason for accidents in almost all countries all over the world. Alcohol Detector in Car project is designed for the safety of the people seating inside the car. Alcohol

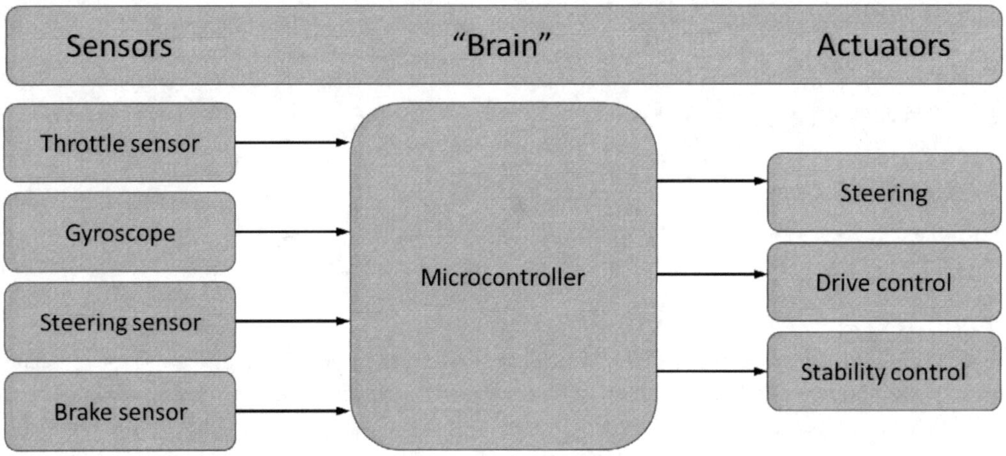

Figure 5. ESP Control schematics of the concept car.

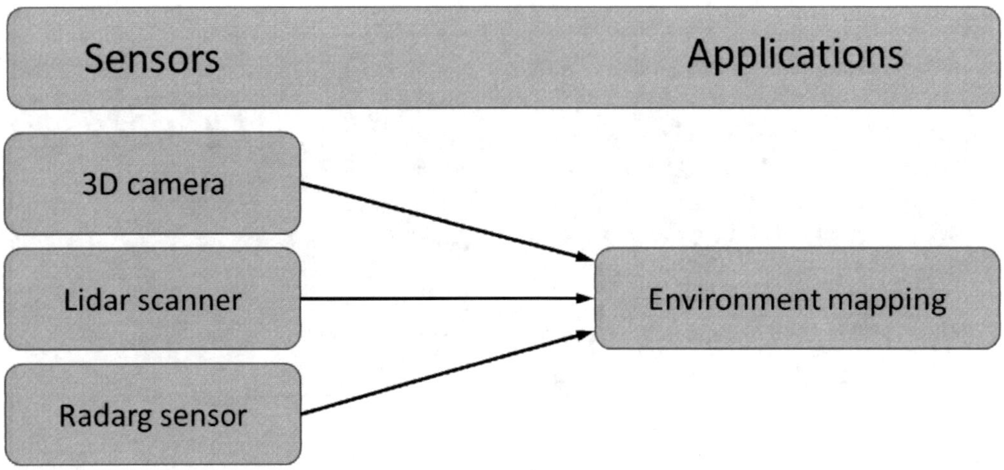

Figure 6. Block diagram of the subsystem.

Detection with Vehicle Controlling project helps to control the vehicle in case the driver has consumed the alcohol. An alcohol breath analyzer project should be fitted / installed inside the vehicle. implementing GSM technology with an alcohol detector. So Alcohol detection & vehicle controlling through text SMS will inform the relatives or owners of the vehicle about the alcohol consumption. implementing GPS technology so that once alcohol detection is done, the system will find out the location of the vehicle. This project is called GPS tracker and alcohol detector with engine locking system using GSM.

Alcohol Detector project can be used in the various vehicles for detecting whether the driver has consumed alcohol or not. Alcohol Detection System in Cars provides an automatic safety system for cars and other vehicles, as well.

4. Mathematical modeling

Mathematical modeling plays an ever increasing role in modern design. It provides a cost effective way to try out different setups and fine tune parameters even before the production of

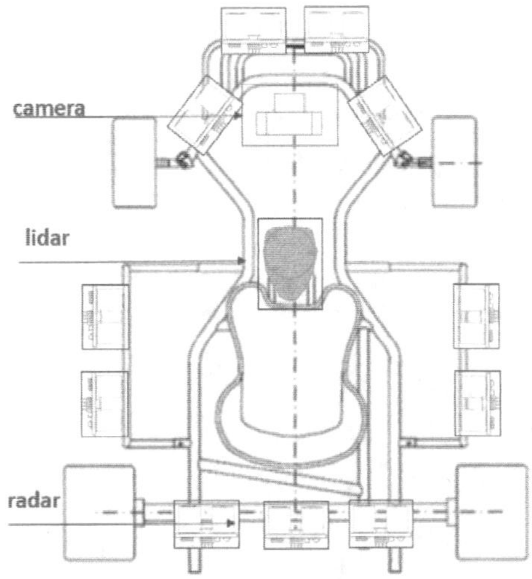

Figure 7. Sensor locations on the concept car.

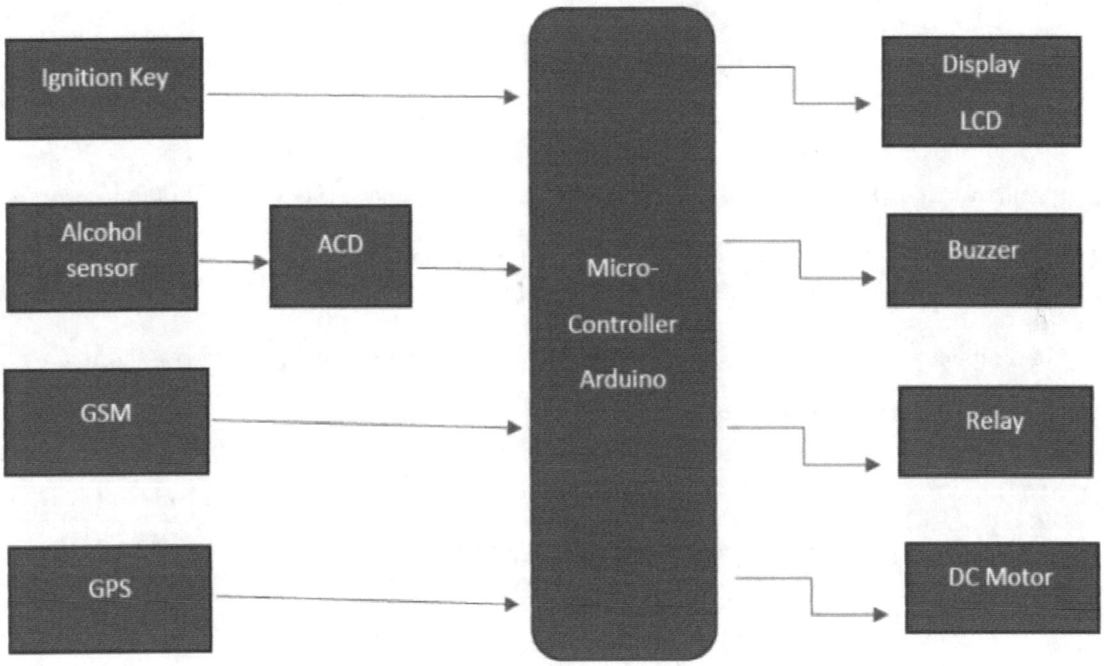

Figure 8. ESP Control schematics of the concept car.

any kind of prototype. Of course, any mathematical model — let it be a formal or a numerical one — has to be compared to a physical realization eventually. Nevertheless, a well-prepared and efficient model can save time, money and other resources. This whole subsection, including the pictures is based on [3].

For this paper we are going to sketch and discuss a general approach to the aforementioned

Table 2. Components.

System components:	
- Controller board	- Display
- Alcohol sensor	- Buzzer
- GSM Shield	- Relay
- GPS Shield	- Electric motor

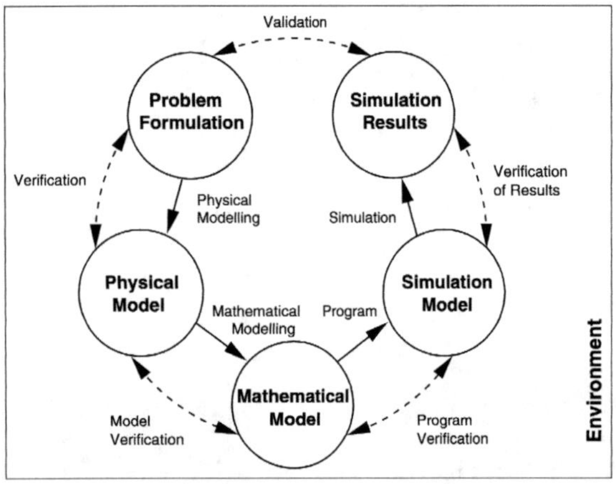

Figure 9. Steps of Mathematical Modeling. (Source: [3])

ESP control. The idea behind the control system is that nowadays the onboard computers are so small but so powerful that they can be used ubiquitously and for quite serious control purposes. They can provide real time assistance or even direct correction in time of need. Let's consider the case of the ESP control system using the following familiar model for wheels depicted in Figure 10.

For simplicity we are going to concentrate on the linear case, where the forces acting on the front and rear wheel are the following, respectively:

$$F_{yF} = C_F \alpha_F, \tag{1}$$

$$F_{yR} = C_R \alpha_R. \tag{2}$$

In equations (1) and (2), the linearized side slip angles of the wheels are

$$\alpha_F = (\delta_F + \Delta\delta_F) - \left(\beta + \frac{\dot{\psi} l_F}{v}\right), \tag{3}$$

$$\alpha_R = \Delta\delta_R - \left(\beta - \frac{\dot{\psi} l_R}{v}\right). \tag{4}$$

Here, besides the steering angle δ_F, we have introduced control through additional steering angles $\Delta\delta_F$ and $\Delta\delta_R$. The resulting linear model looks like this, starting with the state and control vectors:

$$\underline{x}(t) = \begin{bmatrix} \beta \\ \dot{\psi} \end{bmatrix}, \qquad \underline{u}(t) = \begin{bmatrix} \delta_F + \Delta\delta_F \\ \Delta\delta_R \end{bmatrix}. \tag{5}$$

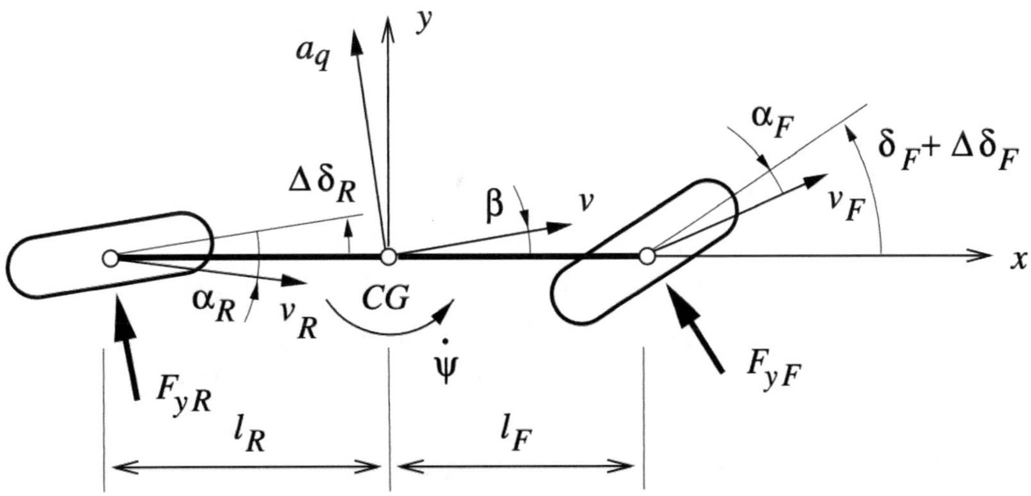

Figure 10. A one-track model of the vehicle. (Source: [3])

The basic linear stat equation with initial value has the usual form:

$$\underline{\dot{x}}(t) = \mathbf{A}\underline{x}(t) + \mathbf{B}\underline{u}(t), \qquad \underline{x}(t_0) = \underline{x}_0, \tag{6}$$

where the matrices **A** and **B** are two-by-two and have the following entries:

$$\begin{aligned} A_{11} &= \frac{C_F + C_R}{mv}, & A_{12} &= -1 - \frac{C_F l_F - C_R l_R}{mv^2}, & B_{11} &= \frac{C_F}{mv}, & B_{12} &= \frac{C_R}{mv}, \\ A_{21} &= \frac{C_F l_F - C_R l_R}{I_z}, & A_{22} &= -\frac{C_F l_F^2 + C_R l_R^2}{I_z v}, & B_{21} &= \frac{C_F l_F}{I_z}, & B_{22} &= \frac{C_R l_R}{I_z}. \end{aligned} \tag{7}$$

Here m is the mass of the car, I_z is the moment of inertia for the z-axis and C_F, C_R are relevant physical parameters of the car. In order to efficiently simulate the system we could use, for example current Multi Body System softwares. To do this we would need to develop an appropriate model. One possible solution is the depicted in Figure 11. For other approaches, see for example [6].

This model combines the input coming from the driver: δ_H and p_d, the steering angle and the brake pedal pressure, respectively, with the information gathered by the sensors. The observer provides estimates for β, the side slip angle of the vehicle, while the controller computes ΔM_{yaw}, the necessary yaw moment correction. The job of the transformer is to calculate the necessary correction for the brake pressure. The Actuator closes the loop and steers the vehicle in the corrected direction.

5. Conclusion

The goal of the enterprise is to help students understand better the issues and challenges that they will face as actual engineers, through a highly educational and entirely hands-on projects. They are going to get familiar with every step on the way of building an actual vehicle. This paper is providing information of the project at initiation state. In the future both the demonstration vehicle will be build, as well as the vehicle drive control, vehicle stability control, and convenience functions. From mathematical point of view our current goal is to understand the already existing and accepted model for all parts of the vehicle. In the future we would like to refine the model in such a way that it suits the emerging phisical car better.

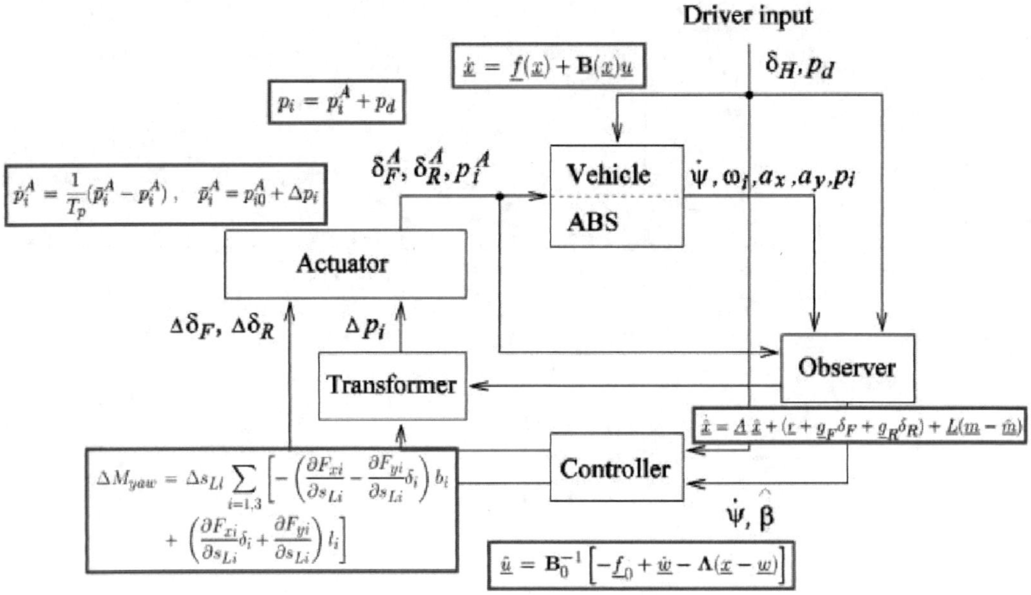

Figure 11. An MBS Model for ESP. (Source: [3])

References

[1] Yedamale, P., 2003. Brushless DC (BLDC) motor fundamentals. Microchip Technology Inc, 20(1), pp.3-15.

[2] Baldursson, S., 2005. Bldc motor modelling and control-a matlab®/simulink® implementation (Master's thesis).

[3] Lugner, Peter, and Manfred Plöchl. "Modelling in vehicle dynamics of automobiles." ZAMM-Journal of Applied Mathematics and Mechanics/Zeitschrift für Angewandte Mathematik und Mechanik: Applied Mathematics and Mechanics 84.4 (2004): 219-236.

[4] Várkonyi-Kóczy, Annamária R. "Intelligent autonomous primary 3D feature extraction in vehicle system dynamics' analysis: Theory and application." Acta Polytechnica Hungarica 5, no. 1 (2008): 5-29.

[5] Vaščák, Ján. "Navigation of mobile robots using potential fields and computational intelligence means." Acta Polytechnica Hungarica 4, no. 1 (2007): 63-74.

[6] Kratmüller, Martin. "The adaptive control of nonlinear systems using the TSK fuzzy logic." Acta Polytechnica Hungarica 6, no. 2 (2009): 5-16.

Modelling Steering Resistance to Save Energy

Zoltán Pusztai[1*], Péter Kőrös[1], Ferenc Friedler[1]

[1]Széchenyi István University, Vehicle Industry Research Center, Egyetem sqr. 1, 9026, Győr, Hungary

*pusztai.zoltan@ga.sze.hu

Abstract. In this paper, the design and vehicle model implementation of a spiral path steering system for a lightweight vehicle is presented. The lateral sliding caused by imperfect rolling in corners means considerable loss, for lightweight vehicles. A spiral path steering mechanism provides an alternative solution for steering generated loss reduction. Theoretically, with the application of a spiral path steering system, ideal Ackerman steering geometry is feasible in all corners. The system is able to replace the common rack and pinion solution in purpose-made vehicles. The main design concepts of spiral path steering are introduced in this article. The described system was realized and installed in an experimental vehicle, where field tests were carried out, to measure the cornering losses during operation. The process of cornering loss measurement is also presented in this paper. The resistance model of cornering is elaborated in a Matlab Simulink environment, based on the measurement results. Vehicle losses are characterized by an extended resistance force model, which can be used by simulations for energy saving purposes. The optimization of vehicle operation can be achieved by simulations, where the corresponding velocity profiles of the vehicle is determined. An accurate steering model is essential for proper vehicle modelling and for the following optimization process.

1. Introduction

Achieving energy efficient vehicle operation is a fundamental goal of all electric vehicles. In urban environments the usage of electric vehicles is more than beneficial, and their CO2 reduction potential is also underestimated. In this paper, a special one seat experimental electric vehicle is investigated, which is dedicated for the Shell Eco-marathon (SEM), energy efficiency competition. The goal of the event is to complete the race in a given time by using the least amount of energy. An energy efficient vehicle operation is a complex task, where driving strategy is acquired by solving the control problem on the suitable vehicle model and objective function. A mathematical vehicle model can be formulated by discrete equations, which provide a robust solution; but in special cases other vehicle model formulation methods could achieve greater accuracy. In the presented work, measurement-based vehicle modelling was used according to the unique characteristics of the vehicle. The extreme low mass, highly aerodynamic shape and optimized drive components of the vehicle results in low overall power demand. The traction force demand is between 10–25 N, which makes high accuracy vehicle modelling reasonable. On the other hand, the arising resistance forces should also be modelled precisely, because even moderate changes could have significant effects on the vehicle's operation. Resistance force modelling can be distinguished in straight moving and in cornering. In this work, the construction of the steering system and the related losses are investigated. [1][2]

IOP Conf. Series: Materials Science and Engineering 1237 (2022) 012016 doi:10.1088/1757-899X/1237/1/012016

Figure 1. SZEmission - experimental electric energy efficient vehicle

2. Ackermann Steering Geometry

The applied steering system and geometry have an important role in vehicle dynamics. In order to achieve energy efficient driving, cornering losses should not be neglected. The ideal Ackerman geometry needs to be maintained in every turn, where tire slip and loss of traction is minimized. The base concept of Ackermann steering is presented on figure 2. The inner and outer wheels are pointing exactly to the actual center of the turn, which is collinear with the rear axle. According to the geometry the inner and outer wheels have different angles. [3][4][5]

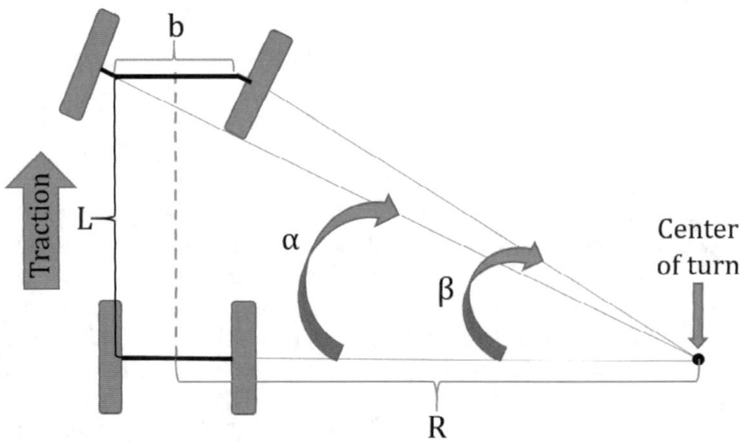

Figure 2. Ackermann steering geometry concept

Perfect rolling in different corners (R) can be achieved by keeping the inner(α) and outer(β) wheels at proper angles as follows:

$$\alpha = tan^{-1}\left(\frac{L}{R+\frac{b}{2}}\right) \tag{1}$$

$$\beta = tan^{-1}\left(\frac{L}{R-\frac{b}{2}}\right) \tag{2}$$

The Ackermann angles depend on the vehicle's characteristics, especially on the kingpin distance in the front suspension. The vehicle parameters of SZEmission are summarized in table 1.

Table 1. Vehicle parameters of SZEmission

Property	Value (mm)
Wheelbase (L)	1300
Front track	1000
Rear track (B)	800
Drag link radius (Ls)	19.5
Kingpin distance (b)	961

Based on equation (1) and (2) and the vehicle parameters in table 1. the discrete Ackermann angles can be determined for the investigated vehicle, which are visualized in figure 3.

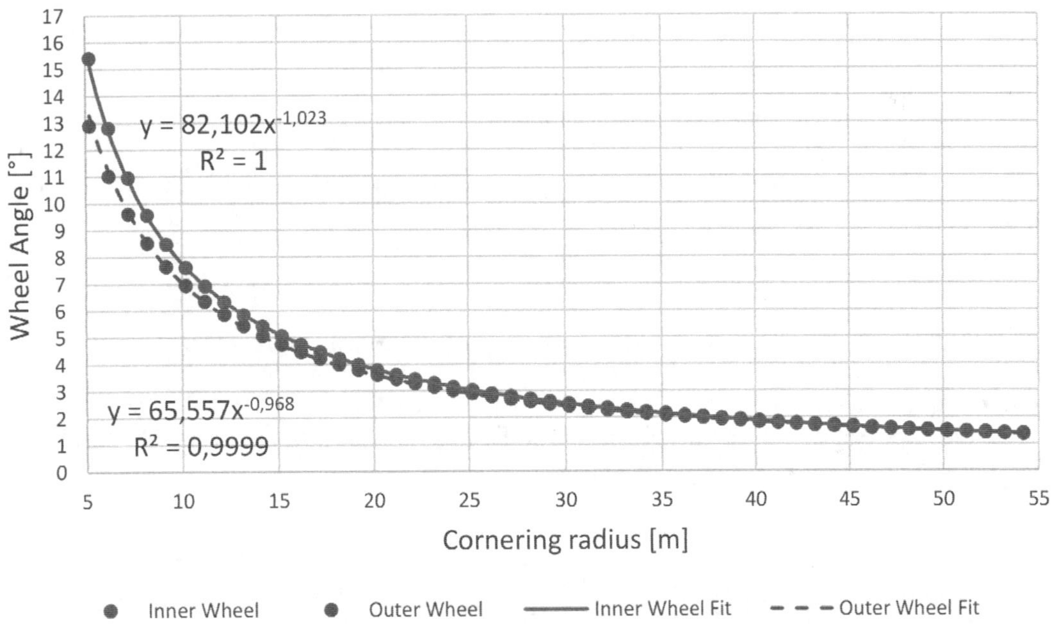

Figure 3. Calculated Ackermann wheel angles of SZEmission

3. Optimal Spiral Path Steering Method

The currently applied steering system is based on the optimal spiral path concept. This concept describes the steering spiral, which guarantees the ideal Ackerman angle in every position of the steering system, without using steer-by-wire methods, which are not allowed by the SEM regulation. In this study, the spiral disk coordinates are determined by the introduced measurement method, instead of CAD modelling. The base concept of measuring is to have a fully assembled suspension, where the displacement of the steering rod (x) can be measured, without the built-in steering system. The aim of measurement is to define the relation between the shaft displacement and wheel angle. In the investigated case, an absolute, guided linear encoder was used to monitor the shaft displacement, while the wheel was turned manually. The schematic layout of the measurement is shown in figure 4. [4][5][6]

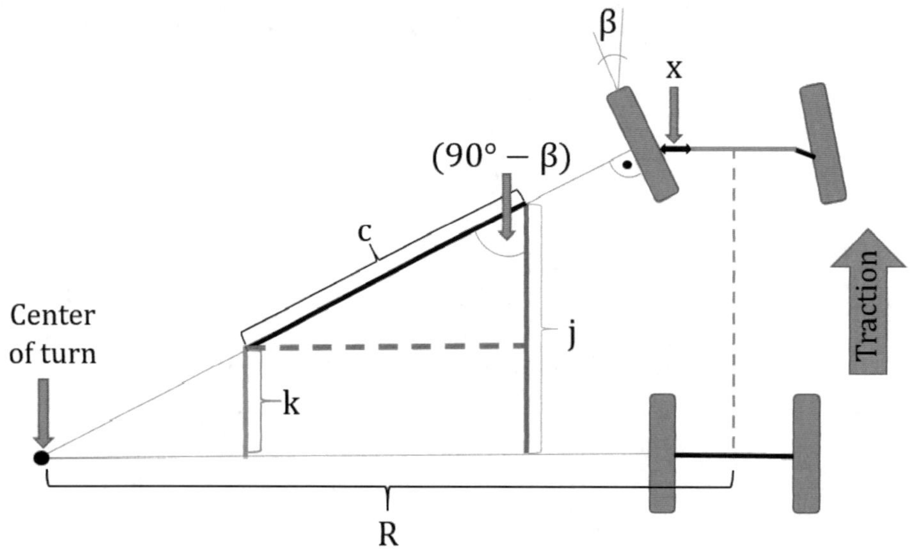

Figure 4. Optimal Spiral Path measuring method

According to the trigonometric relationships in figure 4, the inner wheel angle (β) can be determined as follows:

$$\beta = cos^{-1}\left(\frac{j-k}{c}\right) - 90° \tag{3}$$

For practical reasons the *c* parameter was chosen as a constant value of 1 m, while *k* and *j* distances were measured with laser distance sensors with 0.001 m accuracy. The perpendicularity of *c* section was guaranteed by cross line lasers, which was aligned for every measurement. Several measurements were made, while the wheel angle was manually changed. The results of measurements are presented in figure 5. The measurements were carried out on both sides of the vehicle.

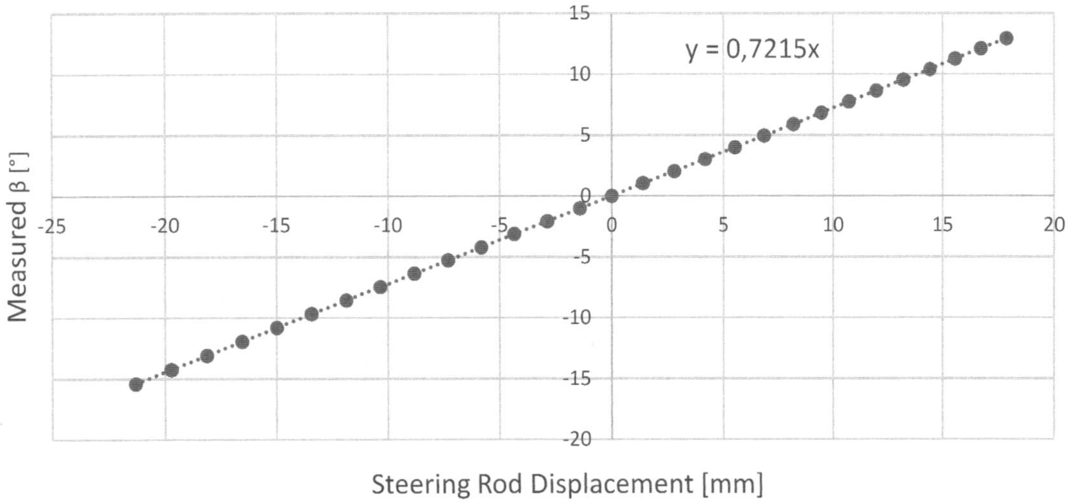

Figure 5. Measured β and steering rod displacement

Based on the measurement results, the steering rod displacements for the theoretical Ackermann angles can be defined. The steering range of the steering wheel was determined from 0°- 160° (λ), which is suitable for an urban vehicle, where the turning ability is important with even with fine steering movement. The base radius (r_{base}) of the spiral path needed to be defined for manufacturability reasons. In the presented work it was considered to be 70 mm.

Table 2. Denotation for spiral path calculation

Property	Denotation
Base radius for spiral path [mm]	r_{base}
Steering rod displacement [mm]	x_{shaft}
Steering angle [°]	λ

The *x* and *y* coordinates of the spiral path can be calculated by using equation (3) and (4) in the whole range of λ.

$$x = (r_{base} + x_{shaft}) * -\cos(\lambda) \quad (4)$$
$$y = (r_{base} + x_{shaft}) * \sin(\lambda) \quad (5)$$

Based on the measurements, calculations need to be done for the left and right wheel, independently. The points can be connected with spline interpolation to form the representative spiral form. The calculated coordinates can be found on figure 6.

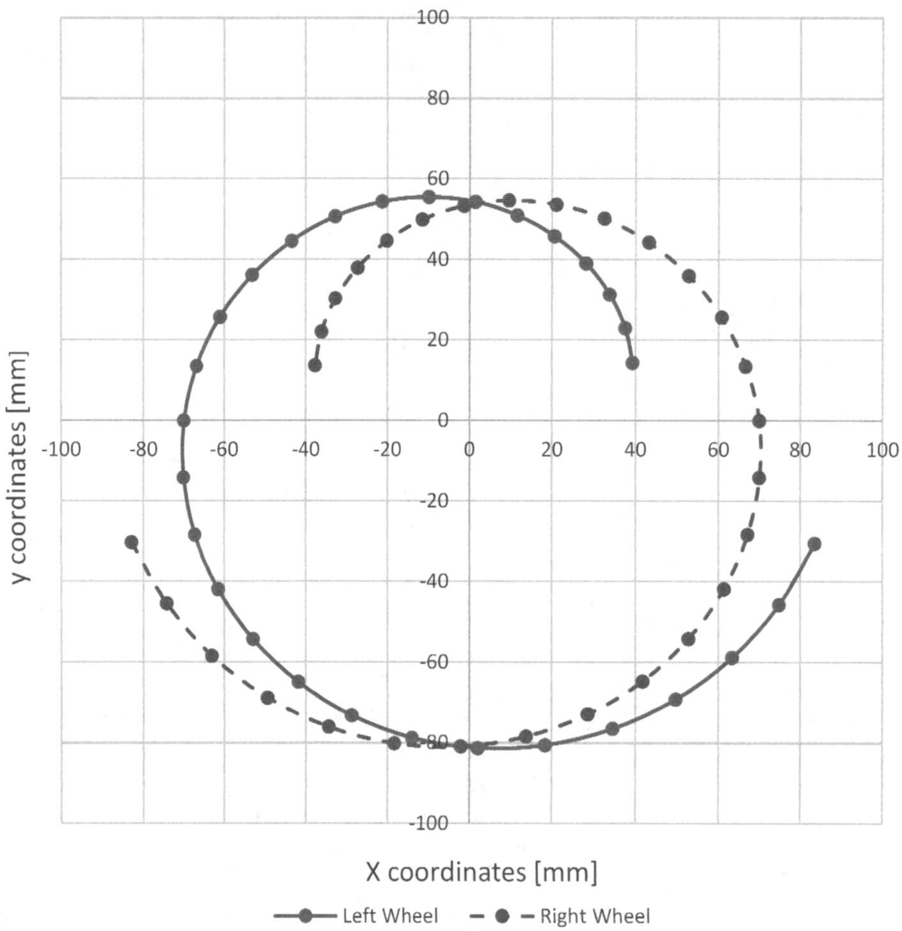

Figure 6. Calculated spiral path coordinates

The vehicle is heading in $y+$ direction and the steering rods are connecting to the spiral path at the points of intersection of the x axis. The left steering rod is connected to the left side (solid line), while the right is connected to the right side (dashed line). The left and right spiral paths are facing in opposite directions to make the connection mechanically feasible. This mechanism allows the wheels to move independently, while keeping the calculated Ackermann angles in every steering position. The spiral paths can be mirrored, if symmetry is assumed. In this study the spiral paths are slightly different because of the combined effect of manufacturing related unwanted asymmetry in the suspension and measurement inaccuracies.

4. Steering System Overview

The calculated spiral disk geometry is implemented in the mechanical system, where the connecting rods are supported by square section linear bearings. An absolute linear encoder, used for the previous measurements, is also placed in the construction, beside the DC stepper motor, both are required for complete autonomous control. According to the rules, the DC stepper motor is not allowed in human driving challenges, so this part of the steering needs to be easily detachable. The drive connection is realized by a synchronous belt drive, with 1:1 gear ratio. The belt drive is also detachable, and the belt pretension can be set in the console of DC stepper motor. The complete assembly of the steering system is presented in figure 7.

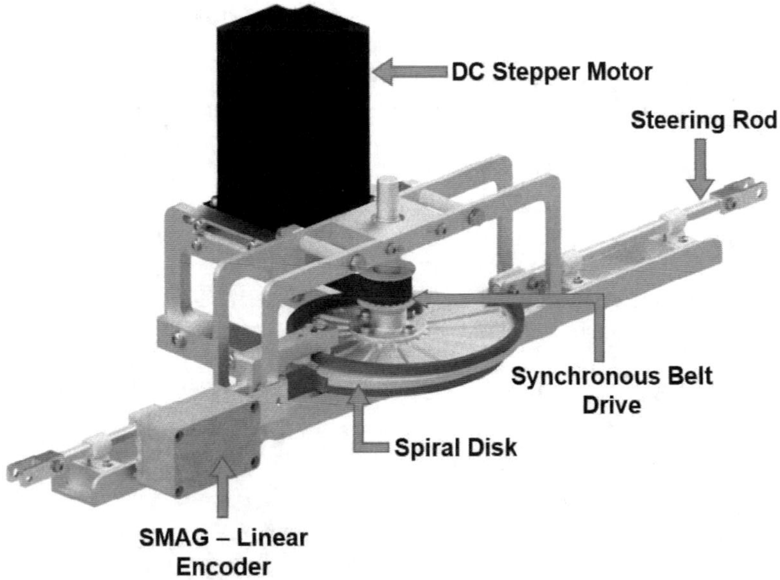

Figure 7. Assembled steering system

The DC stepper is operated based on the commands of the high-level autonomous control system. This control system is responsible for trajectory planning of the vehicle, during autonomous operation. The trajectory planning is based on the evaluation of the data collected from camera and Lidar sensors. The stepper motor rotates the spiral disk, which makes the steering rods move translationally. The translational movement is monitored by the linear encoder providing feedback for the high-level control system. The advantage of this solution is to have accurate feedback, by including all the backlash of the mechanical system. The relation between the linear movement and cornering radius is shown in figure 8. According to this relation, the current cornering radius can be assigned to the steering rod displacement.

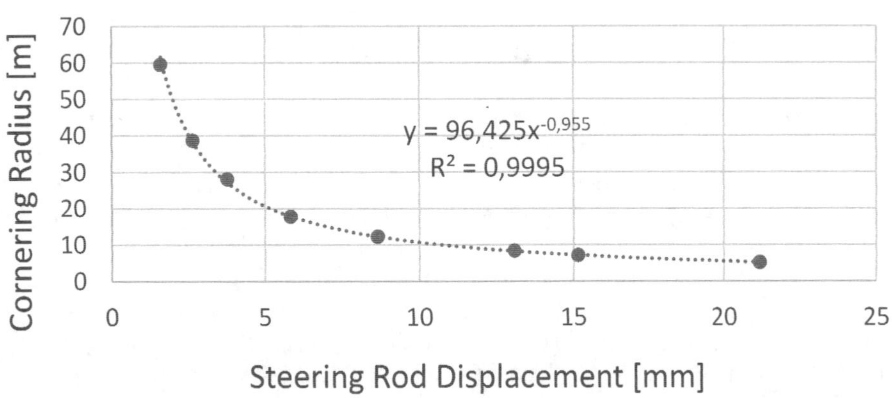

Figure 8. Cornering radius - Steering Rod Displacement

5. Steering Resistance Modelling

The created steering system keeps the wheels in their predetermined position maintaining ideal Ackermann angles, but the cornering still causes more resistance, than moving straight. The cornering losses need to be measured to determine the optimal driving strategy. Field test measurement methodology was introduced to measure cornering losses. Several measuring scenarios were set up with different speed levels and cornering radius values. In these tests, the vehicle speed was controlled by a real-time speed controller, while the driver manually followed the designated route with the given cornering radius. Resistance forces can be calculated from the traction force demand, as it equals to the resistance force, when the system is steady. The results were manually evaluated and combined with the free rolling test to set up the extended resistance force model. The created model was elaborated in a MATLAB environment, where the cumulative resistance forces in cornering can be determined; which includes all the arising resistance of aerodynamics, rolling, and tire slip. The model is shown in figure 9. [2]

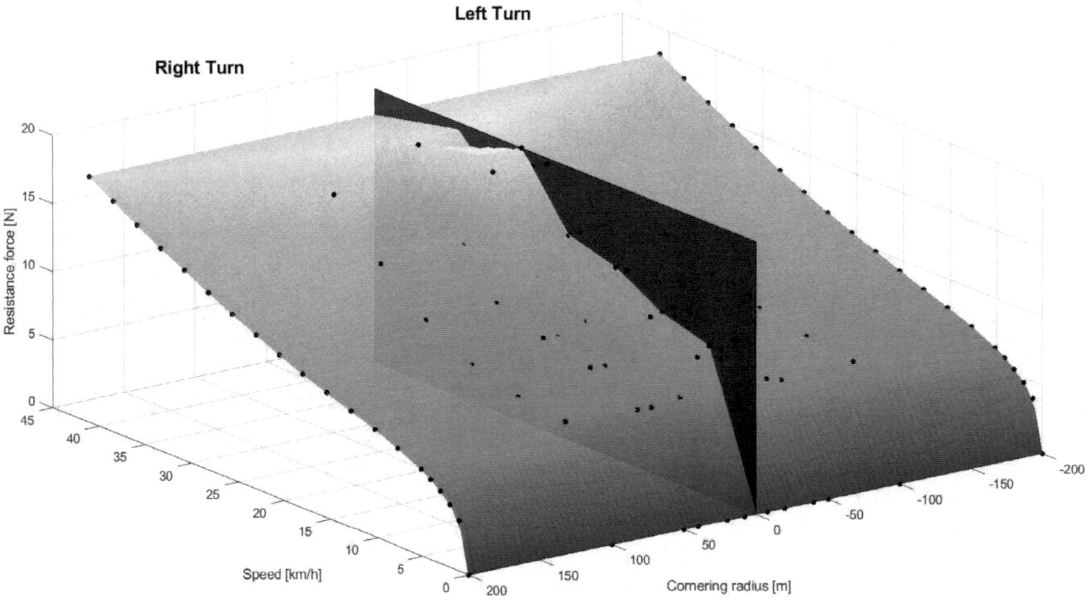

Figure 9. Extended resistance model of the SZEmission

The physical feasibility starts from the cornering radius value of 6 m, and cornering phase lasts until the cornering radius value reaches 200 m, moving straight is considered from 200 m. In figure 9, the black plane distinguishes the left and right turn, and different resistance force values can be found in each side. The powertrain requires more energy in the right corners as the resistance forces are higher at speed levels over 25 km/h. The unequal weight distribution could be responsible for that difference. The rear left wheel is driven by the powertrain, which makes the weight distribution shift to the left. In relatively high speed right corners the rear left wheel is under load and the right wheel could lift of from the ground, decreasing the rolling and the slip of tires. This effect could cause higher resistance force values, which the measurement-based model also confirms. The resistance model is implemented in the vehicle model, where the vehicle speed and vehicle position are determined. Cornering radius can be calculated from the vehicle position on the track and actual vehicle speed, while the grade resistance can also be defined from height and length data of the track. The vehicle is accelerated by the traction force against the summarized resistance force (ΣF_{slow}). The theoretical operation of the track and resistance model is shown in figure 10. The acquired resistance force (ΣF_{slow}) is looped back to the mathematic vehicle model to calculate the vehicle speed and position.

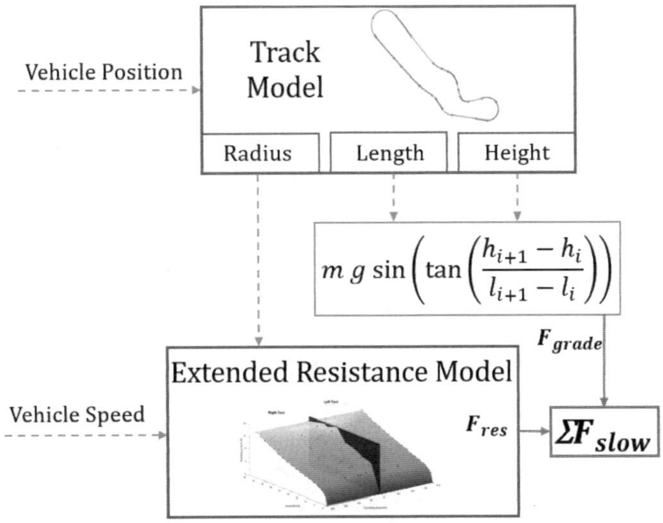

Figure 10. Vehicle model implementation

6. Conclusion

In this paper, a novel approach of steering mechanism design and modelling for lightweight vehicles was presented. A complete process of the measurement based steering design was introduced. The design process is based on static measurements, so the dynamic behavior of the suspension is not taken into consideration. The created steering system was tested in the experimental vehicle, where resistance force measurements were carried out to define energy demand of cornering. The obtained results have shown that the resistance forces are different in the left and right corners because of the vehicle characteristics. The relatively low differences in resistance force (3-5 N) also need to be considered in the vehicle model, while creating driving strategy optimization. The proposed vehicle model implementation is suitable for creating driving strategy optimization.

Acknowledgments

Supported by the ÚNKP-21-3-I-SZE-19 New National Excellence Program of the Ministry for Innovation and Technology from the source of the National Research, Development and Innovation Fund.

NATIONAL RESEARCH, DEVELOPMENT AND INNOVATION OFFICE HUNGARY

Új Nemzeti Kiválóság Program

SZÉCHENYI EGYETEM UNIVERSITY OF GYŐR

References

[1] Hoekstra A., 2019, *The Underestimated Potential of Battery Electric Vehicles to Reduce Emissions*, *Joule*, **3** (6), 1412-14.

[2] Pusztai Z., Kőrös P. and Friedler F., 2021 *Vehicle Model for Driving Strategy Optimization of Energy Efficient Lightweight Vehicle*, Chemical Engineering Transactions, **88**, 6 p, 385-390.

[3] Beckers C., Besselink I. and Nijmeijer H., 2020, *Assessing the impact of cornering losses on the energy consumption of electric city buses*, *Transport and Environment*, **86**, 102360.

[4] Hartani K., Miloud Y. and Miloudi A., 2010, *Electric Vehicle Stability with Rear Electronic Differential Traction, EFEEA'10 International Symposium on Environment Friendly Energies in Electrical Applications*, Ghardaïa, Algeria, 2-4 November 2010

[5] Dollinger, M. and Fischerauer, G., 2021, *Model-Based Range Prediction for Electric Cars and Trucks under Real-World Conditions*. *Energies*, **14**, 5804.

[6] Pusztai Z. and Kőrös P., 2020, *Steering mechanism design for lightweight vehicle*, *GÉP* **3-4, LXXI.**, 5 p, 17-21.

IOP Conf. Series: Materials Science and Engineering 1237 (2022) 012017 doi:10.1088/1757-899X/1237/1/012017

Combining evolutionary optimization with finite element method for optimizing transmission-line tower

Sz Nagy[1], **K Jármai**[2], **A Baksa**[3]

[1] Emerson Automation FCP Kft., Bánki Dónát út 3, Eger 3300, H
[2] Institute of Energy Engineering and Chemical Machinery, University of Miskolc, Egyetemváros, Miskolc 3515, H
[3] Institute of Applied Mechanics, University of Miskolc, Egyetemváros, Miskolc 3515, H

E-mail: szilard.nagy@emerson.com

Abstract. Metaheuristic evolutionary methods are powerful tools for solving non-linear, many dimensional optimisation problems. In this article, one of these methods is combined with finite element method for optimising truss like structure. The advantages of the combined method are presented through a real engineering task, optimisation of transmission line tower. The design of the latter can take a variety of shapes, so in the present case, only the lower part can be optimised. For optimisation, we use the adaptive differential evolution (SaDE) technique. The objective function of the optimisation is the mass of the structure. Finite element analysis is used to determine the stresses. The overall and local buckling constraints have also been considered.

1. Introduction

Meta heuristic and evolutionary algorithms are powerful tools for solving non-linear high dimensional optimisation or searching problems. They are population based stochastic techniques. Self-adaptive differential was proposed in [1] as an improvement of original differential evolution [2]. It was used efficiently in many real cases, such as pattern recognition [3], commonication [4] and engineering [5].

The usage of closed circular hollow section (CHS) in truss like structures is more preferable then other opened sections [6; 7], like angel section. The buckling stiffness of opened sections is very small [5].

In this paper we combine evolutionary optimisation with finite element method for solving optimization of truss like structures from CHS. Proposed general method is used for optimizing real life structure, bottom part of transmission line tower.

2. Constrained self-adaptive differential evolution

Differential evolution (DE) was introduced in 1997 by Storn [2]. It is a simple but powerful evolutionary optimization algorithm. The basic idea behind the algorithm is to produce new individuals by combining positions of randomly selected individuals from the current population. During the iteration steps, it repeats three operations mutation, crossover, and selection until it finally reaches a stopping condition.

Mutation operation produces ${}^{(G)}\mathbf{v}$ mutant vector for each ${}^{(G)}\mathbf{x}$ individual in G^{th} generation population:

- DE/rand/1:

$$ {}^{(G)}\mathbf{v}_i = {}^{(G)}\mathbf{x}_{r_1} + F\left({}^{(G)}\mathbf{x}_{r_2} - {}^{(G)}\mathbf{x}_{r_3}\right) \tag{1} $$

- DE/best/1:

$$ {}^{(G)}\mathbf{v}_i = {}^{(G)}\mathbf{x}_{best} + F\left({}^{(G)}\mathbf{x}_{r_1} - {}^{(G)}\mathbf{x}_{r_2}\right) \tag{2} $$

- DE/current to best/2:

$$ {}^{(G)}\mathbf{v}_i = {}^{(G)}\mathbf{x}_i + F\left({}^{(G)}\mathbf{x}_{best} - {}^{(G)}\mathbf{x}_i\right) + F\left({}^{(G)}\mathbf{x}_{r_1} - {}^{(G)}\mathbf{x}_{r_2}\right) \tag{3} $$

- DE/rand/2:

$$ {}^{(G)}\mathbf{v}_i = {}^{(G)}\mathbf{x}_{r_1} + F\left({}^{(G)}\mathbf{x}_{r_2} - {}^{(G)}\mathbf{x}_{r_3}\right) + F\left({}^{(G)}\mathbf{x}_{r_4} - {}^{(G)}\mathbf{x}_{r_5}\right) \tag{4} $$

- DE/best/2:

$$ {}^{(G)}\mathbf{v}_i = {}^{(G)}\mathbf{x}_{best} + F\left({}^{(G)}\mathbf{x}_{r_1} - {}^{(G)}\mathbf{x}_{r_2}\right) + F\left({}^{(G)}\mathbf{x}_{r_3} - {}^{(G)}\mathbf{x}_{r_4}\right) \tag{5} $$

where i is index of ith individual in population, $r_1 \neq r_2 \neq r_3 \neq r_4 \neq r_5$ are random indices of individuals in population, and finally F is a scaling factor in $[0, 2)$ range.

After mutation, the crossover operation is applied for every ${}^{(G)}\mathbf{v}_i$ and ${}^{(G)}\mathbf{x}_i$ vector pairs to produce ${}^{(G)}\mathbf{u}_i$ trial vectors

$$ {}^{(G)}u_{i,j} = \begin{cases} {}^{(G)}v_{i,j} & \text{if } \mathtt{rand} \leq C_R \vee j_{rand} = j \\ {}^{(G)}x_{i,j} & \text{otherwise} \end{cases} \tag{6} $$

where j is j^{th} coordinate of ith individual, j_{rand} is random coordinate index, and CR is crossover rate.

If the fitness value of newly generated trial vectors better than ${}^{(G)}\mathbf{x}_i$ individual's own, trial vector will replace the original individual during the selection operation

$$ {}^{(G+1)}\mathbf{x}_i = \begin{cases} \mathbf{u}_i & \text{if } \mathcal{F}(\mathbf{u}_i) \leq \mathcal{F}(\mathbf{x}_i) \\ \mathbf{x}_i & \text{otherwise} \end{cases} \tag{7} $$

where $\mathcal{F}()$ is the fitness function to be optimized.

The mutation strategies presented above show different efficiencies in solving different optimization problems. It is not possible to say in advance which method will be the most effective tool for solving a given problem. To achieve a good performance, all five method need to be tried, and F, C_R critical parameters need to be fine tuned. Self-adaptive differential evolution (SaDE) proposes a learning base adaptive method to solve this time-consuming problem in [1].

SaDE algorithm uses two of the five mutation strategies presented above according to equation (1) and (5). In every iteration step, each strategy is applied randomly. The probability of applying of strategy is depend from p variable

$$ {}^{(G)}\mathbf{v}_i = \begin{cases} \text{acc. to eq. (1)} & \text{if } \mathtt{rand} \leq p \\ \text{acc. to eq. (5)} & \text{otherwise} \end{cases} \tag{8} $$

where p probability variable is defined by following

$$p = \frac{n_{s1}\left(n_{s2}+n_{f2}\right)}{n_{s2}\left(n_{s1}+n_{f1}\right)+n_{s1}\left(n_{s2}+n_{f2}\right)} \tag{9}$$

where n_{s1} and n_{s2} are number of trial vectors successfully entering the next generation produced according to equations (2) or (5), n_{f1} and n_{f2} are the number of discarded vectors produced each mutation strategy. The p probability variable is updated every 50th iteration step and at the same time n_{s1}, n_{f1}, n_{s2}, n_{f2} will be reset. In every iteration step, F scaling factor is get a random value in $[0,2)$ range with normal distribution of mean 0.5 and standard deviation 0.3 and it is different for individuals. After every 5th iteration step, C_R crossover rate randomly change in $[0,1)$ range with gaussian distribution of mean C_{Rm} and standard deviation 0.1. C_{Rm} is the average of successfully applied (individual is entering the next generation) C_R values during the learning period, what consists of 25 iteration steps.

Self-adaptive differential evolution is developed for solving continuous optimization problems as a most of evolutionary algorithms. In real life, most of engineering optimization problems are constrained. It must meet mechanical and economical conditions, for example static stress, local buckling, deflection and fatigue. Constrained optimization problem could be written in following general form

$$\begin{array}{rll} \min & f(\mathbf{x}) & \mathbf{x}=\{x_1,x_2,x_3,\ldots,x_n\}\in\mathcal{S}^n \\ \text{subject to} & g_i(\mathbf{x})\le 1 & 1\le i\le r \\ & h_j(\mathbf{x})=0 & 1\le j\le q \end{array} \tag{10}$$

where $\mathbf{x}$ is vector of design variables in n dimensional $\mathcal{S}$ searching space, $g_i(\mathbf{x})$ are r count inequality constraints, $h_j(\mathbf{x})$ are q count equality constraints.

Usage of penalty function is an effective and simple constraint handling method [8]. These methods convert constrained problems to unconstrained problems.

$$\mathcal{F}(\mathbf{x}) = f(\mathbf{x}) + \sum_{i=1}^{r} G_i(\mathbf{x}) + \sum_{j=1}^{q} H_j(\mathbf{x}) \tag{11}$$

$$G_i(\mathbf{x}) = \begin{cases} 0 & \text{if } g_i(\mathbf{x})\le 1 \\ (R_i g_i(\mathbf{x}))^{\beta} & \text{otherwise} \end{cases} \qquad H_j(\mathbf{x}) = \begin{cases} 0 & \text{if } |h_j-\epsilon|\le 0 \\ (Q_j|h_j-\epsilon|)^{\gamma} & \text{otherwise} \end{cases} \tag{12}$$

where $G_i(\mathbf{x})$ and $H_j(\mathbf{x})$ are static penalty functions [9; 10], R_i and Q_j are quite large constants, β and γ are typically 1 or 2 [8].

3. Produce constraints with finite element method

Connections of circular hollow section (CHS) truss structures are frequently modelled as pin connection for elastic analysis [11]. From this property, CHS structures could be modelled with pushed–pulled elements for finite element analysis [12–14], see in figure 1.

Node displacement is only possible along the ξ axis (local coordinate system) through the nodes i and j. In the x–y global coordinate system of whole structure, the projections x, y of this displacement are interpreted.

$${}^{e}\mathbf{u}' = \begin{bmatrix} {}^{e}u_i' & {}^{e}u_j' \end{bmatrix}^T \qquad {}^{e}\mathbf{u} = \begin{bmatrix} {}^{e}u_{ix} & {}^{e}u_{iy} & {}^{e}u_{jx} & {}^{e}u_{jy} \end{bmatrix}^T \tag{13}$$

The connection between the two coordinate systems is the transformation matrix

$${}^{e}\mathbf{T} = \begin{bmatrix} {}^{e}T_{11} & {}^{e}T_{12} & 0 & 0 \\ 0 & 0 & {}^{e}T_{23} & {}^{e}T_{24} \end{bmatrix} \tag{14}$$

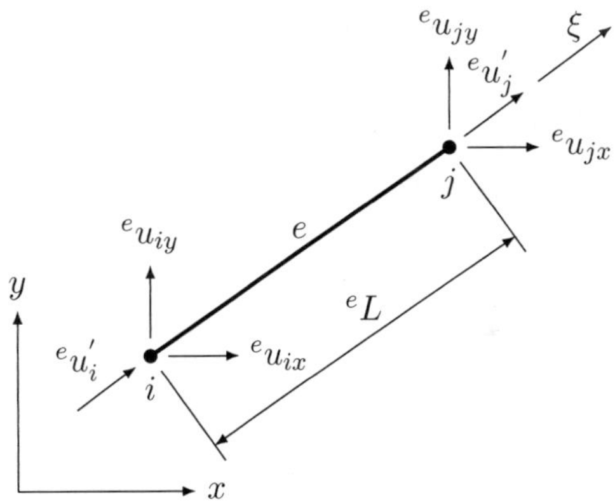

Figure 1. Pushed–pulled truss element.

$$^eT_{11} = ^e T_{23} = \frac{^e u_{jx} - ^e u_{ix}}{^e L} \qquad ^eT_{12} = ^e T_{24} = \frac{^e u_{jy} - ^e u_{iy}}{^e L} \tag{15}$$

$$^e\mathbf{u}' = ^e \mathbf{T} \cdot ^e \mathbf{u} \tag{16}$$

where ^eL is length of eth truss element.

In equilibrium, the Π potential energy of the structure is minimal. This means that the $\delta\Pi$ first variation of the potential energy is zero. The total potential energy of the elements in local and global coordinate system

$$^e\Pi' = \frac{1}{2} \cdot ^e \mathbf{u}'^T \cdot \left(^e\mathbf{K}' \cdot ^e \mathbf{u}' - ^e \mathbf{f}'\right) \tag{17}$$

$$^e\Pi = \frac{1}{2} \cdot \mathbf{u}^T \cdot \left(^e\mathbf{K} \cdot ^e \mathbf{u} - ^e \mathbf{f}\right) \tag{18}$$

where $^e\mathbf{f}'$ and $^e\mathbf{f}'$ are generalized load vectors reduced to nodes in corresponding coordinate system, $^e\mathbf{K}'$ and $^e\mathbf{K}$ are stiffness matrices

$$^e\mathbf{K}' = \frac{^eE \cdot ^e A}{^e L} \begin{bmatrix} 1 & -1 \\ -1 & 1 \end{bmatrix} \tag{19}$$

$$^e\mathbf{K} = ^e \mathbf{T}^T \cdot ^e \mathbf{K} \cdot ^e \mathbf{T} \tag{20}$$

Introducing $\mathbf{u}$ vector of nodal displacements and $\mathbf{f}$ vector of loads

$$\mathbf{u} = \begin{bmatrix} ^1\mathbf{u} & ^2\mathbf{u} & ^3\mathbf{u} & \dots & ^e\mathbf{u} \end{bmatrix}^T \qquad \mathbf{f} = \begin{bmatrix} ^1\mathbf{f} & ^2\mathbf{f} & ^3\mathbf{f} & \dots & ^e\mathbf{f} \end{bmatrix}^T \tag{21}$$

the first variation of potential energy of total structure can be written in the form below

$$\delta\Pi = 0 = \frac{1}{2}\mathbf{u}^T \left(\mathbf{Ku} - \mathbf{f}\right) \tag{22}$$

where $\mathbf{K}$ is stiffens matrix of whole structure according to element connection rules [12]. After applying boundary conditions [12; 13] (fixations, required nodal displacements, etc.) the non-trivial solution of (22) is

$$\mathbf{Ku} = \mathbf{f} \tag{23}$$

For further calculations, it is not enough to know nodal displacements, stress of elements also should be known

$$^e\sigma = \frac{^eE}{^eL} \cdot \left[-^eT_{11} \quad -^eT_{12} \quad ^eT_{11} \quad ^eT_{12}\right] \cdot^e \mathbf{u} \tag{24}$$

First proposed inequality constraint describes resistance of elements to static load

$$g_{Ie} = \begin{cases} \frac{\gamma_{M0}|^e\sigma|}{^e\chi f_y} & \text{if } ^e\sigma < 0 \\ \frac{\gamma_{M0}|^e\sigma|}{f_y} & \text{otherwise} \end{cases} \tag{25}$$

where e means the element under inspection, $^e\chi$ is buckling factor according to [15] by elements, γ_{M0} is generalized safety factor according to [15] and f_y is yield stress.

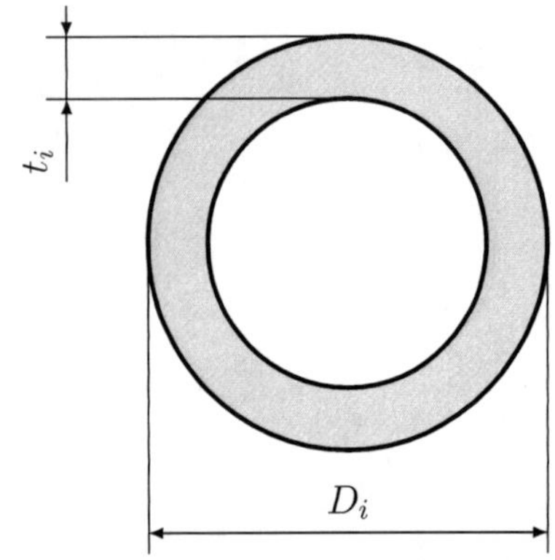

Figure 2. A circular hollow section.

CHS cross section is a thin walled section (see figure 2). which need to be examined for local buckling. The truss like structure may consist of one or more CHS cross sections. Each cross-section of different diameters and wall thicknesses should be inspected for local buckling.

$$g_{IIi} = \frac{D_i f_y}{21150 t_i} \tag{26}$$

Equation (26) is true if only if when unit of f_y is MPa, and unit of D_i, t_i are millimetres.

4. Example truss like structure

Transmission-line towers could be divided into two parts, conical bottom part (see figure 3), and various shaped upper part. In this paper, we have optimized a lower part of an intermediate pylon.

The applied loads were according to [5; 16]. Without detailing of calculations, the size giving load is half wire pulling according to [16]. Forces transmitted from upper part are $F_V = 209.03\,\text{kN}$ vertical force, $F_H = 312.14\,\text{kN}$ horizontal force and $M_h = 2850.5\,\text{kNm}$ bending moment. Forces reduced to nodes with notation of figure 3

$$F_{y1} = \frac{M_h}{2a_2} - \frac{F_V}{4} = \frac{2850.5\,\text{kNm}}{2 \cdot 3.7\,\text{m}} - \frac{209.03\,\text{kN}}{4} = 332.94\,\text{kN} \tag{27}$$

$$F_{y1} = \frac{M_h}{2a_2} + \frac{F_V}{4} = \frac{2850.5\,\text{kNm}}{2 \cdot 3.7\,\text{m}} + \frac{209.03\,\text{kN}}{4} = 437.46\,\text{kN} \tag{28}$$

where $a_2 = 3.7\,\text{m}$ is the width of upper pylon part. It is sufficient to perform the calculations on only one inclined plane relevant to the loads as shown in figure 4. In the case of a pyramid with a side skew of $\beta = 80°$, the forces acting on the plane

$$F_1 = \frac{F_{y1}}{\sin\beta} = \frac{332,94kN}{\sin 80°} = 338,08kN \tag{29}$$

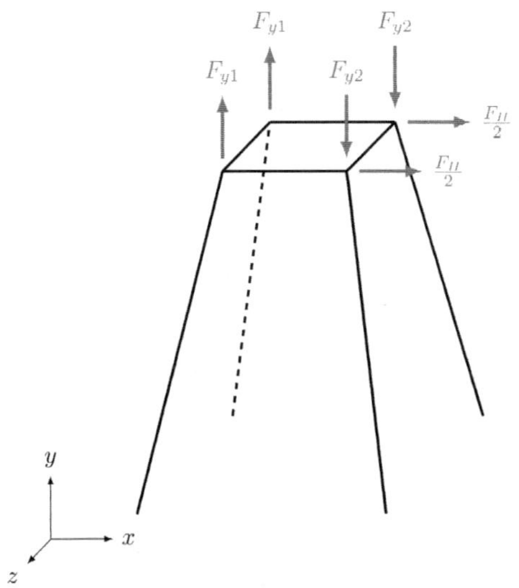

Figure 3. Simplified illustration of lower part of transmission line tower.

$$F_2 = \frac{F_{y2}}{\sin\beta} = \frac{437,46kN}{\sin 80°} = 444,21kN \tag{30}$$

$$F_3 = \frac{F_H}{2} = \frac{312,14kN}{2} = 156,07kN \tag{31}$$

Trusses of structure were divided into three cross-section group. The first group is elements of side column (1–10 elements). Trusses in second group are horizontal bars (17–22 elements).

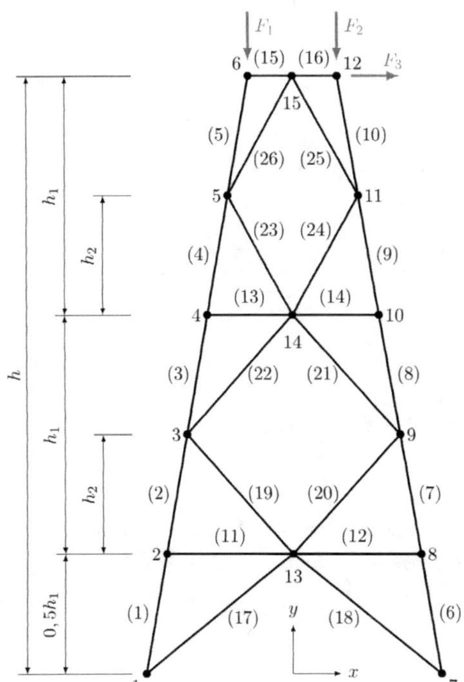

Figure 4. Planar truss like structure.

The final group consists of grid elements forming a deltoid and bottom triangle (23–26). In each cross-section group, CHS is used. The characteristic dimensions according to figure 2 were outer diameters D_1, D_2, D_3, and wall thicknesses t_1, t_2, t_3.

Weight minimum was the target function of optimisation

$$f(\mathbf{x}) = \sum_{e=1}^{n_e} \frac{\pi \rho A_e^2}{4} \tag{32}$$

where ρ is density of steel, A_e is cross-sectional area of eth elements, n_e is number of elements, $\mathbf{x} = [D_1, D_2, D_3, t_1, t_2, t_3]^T$ is vector of unknowns. According to (11), (25), (26) and (32) the final fitness function for SaDE algorithm

$$\mathcal{F}(\mathbf{x}) = f(\mathbf{x}) + \sum_{e=1}^{n_e} G_e\left(g_{Ie}(\mathbf{x})\right) + \sum_{i=1}^{3} G_i\left(g_{IIi}(\mathbf{x})\right) \tag{33}$$

$R_e = R_i = 10^6$ penalty constants was used for $G_e(\mathbf{x})$ and $G_i(\mathbf{x})$ penalty functions.

5. Result of optimisation

Several different topologies with different number of grids were examined. Grid division was changed between 1–6. Also yield stress of raw material steel was changed between $f_y = 235\,\text{MPa}$ and $f_y = 690\,\text{MPa}$. You can see results in figure 5 and figure 6.

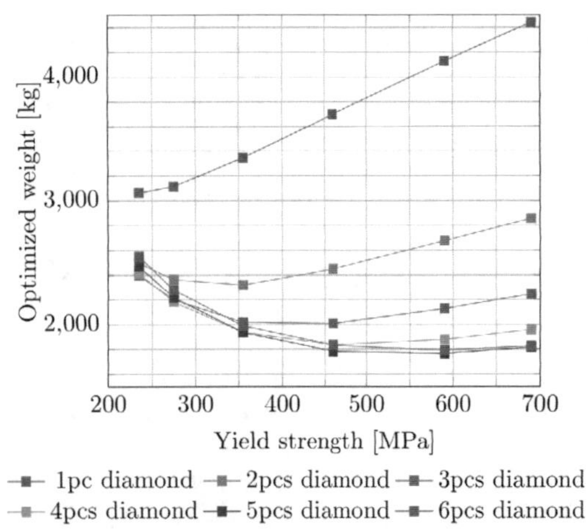

Figure 5. Optimized weight with different grid division and raw material

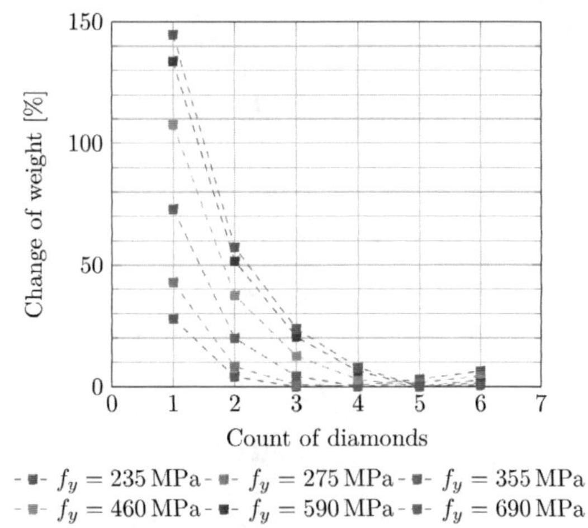

Figure 6. Percentage change in optimised weight

Minimal weight could be found when number of grid division was 5 and f_y yield stress was $460MPa$. As you can see the optimized weight decreases if count of grid division increase. Usage of higher strength steel does not result lighter structure. This is because a thicker bar is required for greater yield strength due to buckling.

6. Conclusion

We have combined an evolutionary optimisation algorithm – such as self-adaptive differential evolution – with finite element method for optimizing truss like structures from circular hollow sections. The proposed method could be used for fully automatic optimization, when finite element model of structure is known.

We have optimised a bottom part of an intermediate transmission-line tower with this method. The optimized weight not necessary decrease if higher strength steel is used. On one hand higher yield stress increases allowable static stress, but on the other hand it is harmful for χ buckling factor.

References

[1] Qin A and Suganthan P 2005 Self-adaptive differential evolution algorithm for numerical optimization *IEEE Congress on Evolutionary Computation* vol 2 pp 1785–1791

[2] Storn R and Price K 1997 *Journal of Global Optimization* **11** 341–359

[3] Ilonen J, Kamarainen J K and Lampinen J 2003 *Neural Processing Letters* **17** 93–105

[4] Nurhan K 2005 *Journal on Applied Signal Processing* **8** 1269–1276

[5] Károly J and József F *Innovative design of steel structures (in Hungarian)* (Miskolc: Gazdász Elasztik Kiadó és Nyomda) ISBN 978-963-358-064-6

[6] Rao G 1995 *Computers & Structures* **57** 81–92

[7] da Silva J, Vellasco P d S, de Andrade S and de Oliveira M 2005 *Journal of Constructional Steel Research* **61** 1108–1134

[8] Yeniay O 2005 *Mathematical and Computational Applications* **10** 45–56 ISSN 2297-8747

[9] Homaifar A, Qi C X and Lai S H 1994 *Simulation* **62** 242–254

[10] Morales A K and Quezada C V 1998 A universal eclectic genetic algorithm for constrained optimization *Proceedings 6th European Congress on Intelligent Techniques and Soft Computing (EUFIT)* vol 2 (Verlag) pp 518–522

[11] Wardenier J, Kurobane Y, Packer J A, van der Vegte A and Zhao X L 2008 *Design guide for circular hollow section (CHS) joints under predominantly static loading* (Zürich: CIDECT)

[12] István P 1999 *Finite element methods in enginerring practice (in Hungarian)* (Miskolc-Egyetemváros: Miskolci Egyetemi Kiadó) ISBN 963-661-312-5

[13] António J M F *MATLAB Codes for Finite Element Analysis* (Berlin: Springer-Verlag) ISBN 978-94-007-8955-5

[14] Ian M S and Lee M *Programming the Finite Element Method, 4th edition* (West Sussex: John Wiley and Sons Ltd.) ISBN 0-470-84969-3

[15] EN 1993-1-1:2009 Eurocode 3: Design of steel structures - Part 1-1: General rules and rules for buildings

[16] MSZ 151-1:2000 Power overhead lines. Installation requirements for overhead lines with nominal voltage greater than 1 kV

Acknowledgments

The research was supported by the Hungarian National Research, Development and Innovation Office under the project number K 134358, and by the NTP-SZKOLL-20-0022 identifier "Focus'21-Focus on community by developing digital competencies" project, supported by the Ministry of Human Resources and Human Resources Support Manager.

AUTHOR INDEX